Analysis and
Mathematical Physics

LTCC Advanced Mathematics Series

Series Editors: Shaun Bullett *(Queen Mary University of London, UK)*
Tom Fearn *(University College London, UK)*
Frank Smith *(University College London, UK)*

Published

Forthcoming

LTCC Advanced Mathematics Series — Volume 6

Analysis and Mathematical Physics

Editors

Shaun Bullett
Queen Mary University of London, UK

Tom Fearn
University College London, UK

Frank Smith
University College London, UK

World Scientific

NEW JERSEY • LONDON • SINGAPORE • BEIJING • SHANGHAI • HONG KONG • TAIPEI • CHENNAI • TOKYO

Published by

World Scientific Publishing Europe Ltd.

57 Shelton Street, Covent Garden, London WC2H 9HE

Head office: 5 Toh Tuck Link, Singapore 596224

USA office: 27 Warren Street, Suite 401-402, Hackensack, NJ 07601

Library of Congress Cataloging-in-Publication Data

Names: Bullett, Shaun, 1947–   editor. | Fearn, T., 1949–   editor. |
    Smith, F. T. (Frank T.), 1948–   editor.
Title: Analysis and mathematical physics / edited by Shaun Bullett (Queen Mary University of
    London, UK), Tom Fearn (University College London, UK),
    Frank Smith (University College London, UK).
Other titles: Analysis and mathematical physics (Hackensack, N.J.)
Description: [Hackensack, N.J.] : World Scientific, 2017. | Series: LTCC advanced
    mathematics series ; volume 6 | Includes bibliographical references.
Identifiers: LCCN 2016036916 | ISBN 9781786340986 (hc : alk. paper)
Subjects: LCSH: Mathematical analysis. | Mathematical physics. | Geometry, Differential.
Classification: LCC QC20.7.A5 A534 2017 | DDC 530.15--dc23
LC record available at https://lccn.loc.gov/2016036916

British Library Cataloguing-in-Publication Data
A catalogue record for this book is available from the British Library.

Desk Editors: V. Vishnu Mohan/Mary Simpson

Typeset by Stallion Press
Email: enquiries@stallionpress.com

Chapter 2 was contributed by Professor Yuri Safarov. Yuri was highly regarded as a teacher, researcher and colleague. We dedicate this volume as a tribute to his memory.

Preface

The London Taught Course Centre (LTCC) for PhD students in the Mathematical Sciences has the objective of introducing research students to a broad range of topics. For some students, some of these topics might be of obvious relevance to their PhD projects, but the relevance of most will be much less obvious or apparently non-existent. However all of us involved in mathematical research have experienced that extraordinary moment when the penny drops and some tiny gem of information from outside ones immediate research field turns out to be the key to unravelling a seemingly insoluble problem, or to opening up a new vista of mathematical structure. By offering our students advanced introductions to a range of different areas of mathematics, we hope to open their eyes to new possibilities that they might not otherwise encounter.

Each volume in this series consists of chapters on a group of related themes, based on modules taught at the LTCC by their authors. These modules were already short (five two-hour lectures) and in most cases the lecture notes here are even shorter, covering perhaps three-quarters of the content of the original LTCC course. This brevity was quite deliberate on the part of the editors — we asked the authors to confine themselves to around 35 pages in each chapter, in order to allow as many topics as possible to be included in each volume, while keeping the volumes digestible. The chapters are "advanced introductions", and readers who wish to learn more are encouraged to continue elsewhere. There has been no attempt to make the coverage of topics comprehensive. That would be impossible in any case — any book or series of books which included all that a PhD student in mathematics might need to know would be so large as to be totally unreadable. Instead what we present in this series is a cross-section of some of the topics, both classical and new, that have appeared in LTCC modules in the nine years since it was founded.

The present volume covers topics in analysis and mathematical physics. The main readers are likely to be graduate students and more experienced researchers in the mathematical sciences, looking for introductions to areas with which they are unfamiliar. The mathematics presented is intended to be accessible to first year PhD students, whatever their specialised areas of research. Whatever your mathematical background, we encourage you to dive in, and we hope that you will enjoy the experience of widening your mathematical knowledge by reading these concise introductory accounts written by experts at the forefront of current research.

Shaun Bullett, Tom Fearn, Frank Smith

Contents

Chapter 1

Differential Geometry and Mathematical Physics

Andrew Hone and Steffen Krusch

School of Mathematics, Statistics & Actuarial Science
University of Kent, Canterbury CT2 7NF, UK

The chapter will illustrate how concepts in differential geometry arise naturally in different areas of mathematical physics. We will describe manifolds, fibre bundles, (co)tangent bundles, metrics and symplectic structures, and their applications to Lagrangian mechanics, field theory and Hamiltonian systems, including various examples related to integrable systems and topological solitons.

1. Manifolds

Manifolds are a central concept in mathematics and have natural applications to problems in physics. Here we provide an example-led introduction to manifolds and introduce important additional structures.

A simple, yet non-trivial, example of a manifold is the 2-sphere S^2, given by the set of points

$$S^2 = \left\{ (x_1, x_2, x_3) : \sum_{i=1}^{3} x_i^2 = 1 \right\}.$$

Often, we label the points on the sphere by polar coordinates:

$$x_1 = \cos\phi \sin\theta, \quad x_2 = \sin\phi \sin\theta, \quad x_3 = \cos\theta,$$

where $0 \leq \phi < 2\pi$, $0 \leq \theta \leq \pi$.

However, there is the following problem: we cannot label S^2 with a single coordinate system such that nearby points have nearby coordinates, and every point has unique coordinates.

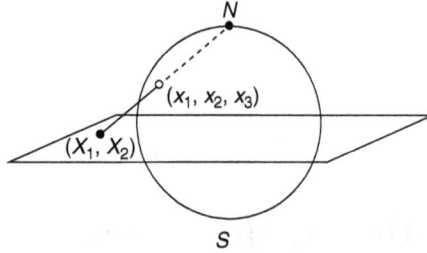

Fig. 1.   Stereographic projection.

Polar coordinates suffer both problems: the point $(1,0,0)$ has coordinates $\theta = \pi/2$ and $\phi = 0$, whereas the nearby point $(\sqrt{1-\epsilon^2}, -\epsilon, 0)$ for small ϵ has coordinates $\theta = 0$ and $\phi \approx 2\pi - \epsilon$, which is not nearby. Furthermore, the north pole $(0,0,1)$ is given by $\theta = 0$ and ϕ is arbitrary, which is clearly not unique. An alternative is to use stereographic coordinates which are constructed as follows. Let N be the north pole given by $(0,0,1)$. Given any point $P \in S^2 \setminus \{N\}$, draw a line through N and P. This line intersects the x_1x_2-plane at the point $(X_1, X_2, 0)$, see Fig. 1. A short calculation shows that these coordinates are given by $X_1 = \frac{x_1}{1-x_3}, X_2 = \frac{x_2}{1-x_3}$. Stereographic coordinates give every point other than the North pole N a unique set of coordinates. However, close to the North pole nearby points do not have nearby coordinates. In order to proceed, we have to use more than one coordinate system, and this gives rise to the definition of a manifold.

Definition 1.1. M is an m-dimensional (differentiable) manifold if

- M is a topological space.
- M comes with family of charts $\{(U_i, \phi_i)\}$ known as an *atlas*.[a]
- $\{U_i\}$ is family of open sets covering M: $\bigcup_i U_i = M$.
- ϕ_i is homeomorphism from U_i onto an open subset U_i' of $\mathbb{R}^m$.
- For each $U_i \cap U_j \neq \emptyset$, the map

$$\psi_{ij} = \phi_i \circ \phi_j^{-1} : \phi_j(U_i \cap U_j) \to \phi_i(U_i \cap U_j)$$

is C^∞. The ψ_{ij} are called *crossover maps*.

The above definition is illustrated in Fig. 2.

---

[a]Two atlases are *compatible* if their union is again an atlas. Compatibility introduces an equivalence relation with equivalence classes known as *differential structures*.

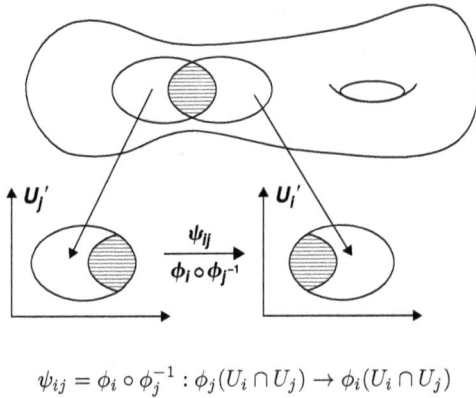

$$\psi_{ij} = \phi_i \circ \phi_j^{-1} : \phi_j(U_i \cap U_j) \to \phi_i(U_i \cap U_j)$$

Fig. 2. A manifold and two overlapping charts.

Example 1.2 (The 2-sphere S^2). The projection from the North pole gives

$$X_1 = \frac{x_1}{1 - x_3}, \quad X_2 = \frac{x_2}{1 - x_3},$$

with open sets $U_1 = S^2 \setminus \{N\}$ and $U_1' = \mathbb{R}^2$. Then the first chart is given by

$$\phi_1 : U_1 \to \mathbb{R}^2$$
$$(x_1, x_2, x_3) \mapsto (X_1, X_2).$$

Similarly, the projection from the South pole results in

$$Y_1 = \frac{x_1}{1 + x_3}, \quad Y_2 = \frac{x_2}{1 + x_3},$$

with open sets $U_2 = S^2 \setminus \{S\}$ and $U_2' = \mathbb{R}^2$. Then the second chart is

$$\phi_2 : U_2 \to \mathbb{R}^2$$
$$(x_1, x_2, x_3) \mapsto (Y_1, Y_2).$$

The *crossover map* $\psi_{21} = \phi_2 \circ \phi_1^{-1}$ can then be calculated as

$$\psi_{21} : \mathbb{R}^2 \setminus \{0\} \to \mathbb{R}^2; (X_1, X_2) \mapsto (Y_1, Y_2) = \left(\frac{X_1}{X_1^2 + X_2^2}, \frac{-X_2}{X_1^2 + X_2^2} \right). \quad (1.1)$$

Example 1.3 (The 3-sphere S^3). The 3-sphere is given by the set of points

$$S^3 = \left\{ (x_1, x_2, x_3, x_4) : \sum_{i=1}^{4} x_i^2 = 1 \right\}.$$

It can also be labelled by stereographic coordinates, which work the same way as for S^2, e.g. $X_i = x_i/(1-x_4)$ for $i = 1, 2, 3$ define the projection from the North pole N. However, S^3 can also be identified with the Lie group $SU(2)$, i.e. complex 2×2 matrices which satisfy

$$U\,U^\dagger = U^\dagger U = 1 \quad \text{and} \quad \det U = 1. \tag{1.2}$$

Upon setting

$$U = \begin{pmatrix} z_1 & z_2 \\ -\bar{z}_2 & \bar{z}_1 \end{pmatrix},$$

all the conditions in (1.2) are seen to be satisfied provided $|z_1|^2 + |z_2|^2 = 1$, which corresponds to the equation for S^3 e.g. if we set $z_1 = x_1 + ix_2$, $z_2 = x_3 + ix_4$. Hence the 3-sphere can be given a group structure.

1.1. *Maps between manifolds*

Let M be an m-dimensional manifold with charts $\phi_i : U_i \to \mathbb{R}^m$ and N be an n-dimensional manifold with charts $\psi_j : \tilde{U}_j \to \mathbb{R}^n$. Let $f : M \to N$ be a map between the two manifolds, so a point $p \mapsto f(p)$. This has a coordinate presentation

$$F_{ji} = \psi_j \circ f \circ \phi_i^{-1} : \mathbb{R}^m \to \mathbb{R}^n, x \mapsto \psi_j(f(\phi_i^{-1}(x))),$$

where $x = \phi_i(p)$ ($p \in U_i$ and $f(p) \in \tilde{U}_j$). Using the coordinate presentation, all the rules of calculus in $\mathbb{R}^n$ work for maps between manifolds. If the presentations F_{ji} are differentiable in all charts, then f is said to be differentiable.

2. Fibre Bundles

Often manifolds can be built up from smaller manifolds. An important example is the Cartesian product of two manifolds, $E = M \times F$. *Fibre bundles* are manifolds which look like Cartesian products *locally*, but not *globally*. This concept is very useful for physics: non-trivial fibre bundles occur in general relativity, and also in field theories due to boundary conditions "at infinity."

Definition 2.1. A *fibre bundle* (E, π, M, F, G) consists of the following properties:

- A manifold E called the *total space*, a manifold M called the *base space* and a manifold F called the *fibre* (or typical fibre).
- A surjection $\pi : E \to M$ called the *projection*. The inverse image of a point $p \in M$ is called the fibre at p, namely $\pi^{-1}(p) = F_p \cong F$.
- A Lie group G called the *structure group* which acts on F on the left.
- A collection of open sets $\{U_i\}$ covering M and diffeomorphisms $\phi_i : U_i \times F \to \pi^{-1}(U_i)$, such that $\pi \circ \phi_i(p, f) = p$. Each map ϕ_i is called a *local trivialization*, since ϕ_i^{-1} maps $\pi^{-1}(U_i)$ to $U_i \times F$.
- Transition functions $t_{ij} : U_i \cap U_j \to G$, such that $\phi_j(p, f) = \phi_i(p, t_{ij}(p)f)$, so $t_{ij} = \phi_i^{-1} \circ \phi_j$ holds for fixed p.

The transition functions t_{ij} satisfy the conditions

$$
\begin{aligned}
t_{ii}(p) &= e & \forall p \in U_i, \\
t_{ij}(p) &= t_{ji}^{-1}(p) & \forall p \in U_i \cap U_j, \\
t_{ij}(p) \cdot t_{jk}(p) &= t_{ik}(p) & \forall p \in U_i \cap U_j \cap U_k,
\end{aligned}
$$

to be consistent with the fact that $t_{ij} \in G$. If all the transition functions are the identity map e, then the fibre bundle is the *trivial bundle* $E = M \times F$. In general, the transition functions of two local trivializations $\{\phi_i\}$ and $\{\tilde{\phi}_i\}$ for fixed $\{U_i\}$ are related via

$$\tilde{t}_{ij}(p) = g_i^{-1}(p) \cdot t_{ij}(p) \cdot g_j(p),$$

where for fixed p, we define $g_i : F \to F$, $g_i = \phi_i^{-1} \circ \tilde{\phi}_i$. For the trivial bundle, $t_{ij}(p) = g_i^{-1}(p) \cdot g_j(p)$.

Example 2.2 ($U(1)$ bundles over $\mathbf{S}^2$ and magnetic monopoles). Consider a fibre bundle with fibre $U(1)$ and base space S^2. Let $\{U_N, U_S\}$ be an open covering of S^2 where

$$U_N = \{(\theta, \phi) : 0 \le \theta < \pi/2 + \epsilon, 0 \le \phi < 2\pi\},$$

$$U_S = \{(\theta, \phi) : \pi/2 - \epsilon < \theta \le \pi, 0 \le \phi < 2\pi\}.$$

The intersection $U_N \cap U_S$ is a strip which is basically the equator. The local trivializations are $\phi_N^{-1}(u) = (p, e^{i\alpha_N})$, $\phi_S^{-1}(u) = (p, e^{i\alpha_S})$, where $p = \pi(u)$.

Possible transition functions are $t_{NS} = e^{in\phi}$, where $n \in \mathbb{Z}$, and define differ-ent bundles, denoted P_n. The fibre coordinates in $U_N \cap U_S$ are related via

$$e^{i\alpha_N} = e^{in\phi}e^{i\alpha_S}.$$

If $n = 0$ this is the trivial bundle $P_0 = S^2 \times S^1$. For $n \neq 0$ the $U(1)$ bundle P_n is twisted. P_n is an example of a *principal bundle* because the fibre is the same as the structure group $G = U(1)$. In physics, P_n is interpreted as a magnetic monopole of charge n.

Example 2.3 (Hopf bundle). The Hopf map $\pi : S^3 \rightarrow S^2$, with $S^3 = \{x \in \mathbb{R}^4 : x_1^2 + x_2^2 + x_3^2 + x_4^2 = 1\}$, is defined by

$$\xi_1 = 2(x_1 x_3 + x_2 x_4), \quad \xi_2 = 2(x_2 x_3 - x_1 x_4), \quad \xi_3 = x_1^2 + x_2^2 - x_3^2 - x_4^2,$$

which implies $\xi_1^2 + \xi_2^2 + \xi_3^2 = 1$. It turns out that with this choice of coor-dinates S^3 can be identified with P_1, a non-trivial $U(1)$ bundle over S^2, known as the Hopf bundle.

2.1. *Tangent vectors*

Given a curve $c : (-\epsilon, \epsilon) \rightarrow M$ and a function $f : M \rightarrow \mathbb{R}$, we define the tangent vector $X[f]$ at $c(0)$ by taking the the directional derivative of $f(c(t))$ along $c(t)$ at $t = 0$, namely,[b]

$$X[f] = \left.\frac{df(c(t))}{dt}\right|_{t=0}.$$

In local coordinates (x^μ), with $n = \dim M$, this becomes

$$\left.\sum_{\mu=1}^{n} \frac{\partial f}{\partial x^\mu} \frac{dx^\mu(c(t))}{dt}\right|_{t=0}.$$

Hence tangent vectors act on functions via

$$X[f] = X^\mu \left(\frac{\partial f}{\partial x^\mu}\right),$$

where above there is an implicit sum over repeated indices. (This is the Einstein summation convention, and henceforth we shall always use this.)

---

[b] Alternatively, tangent vectors can be defined as equivalence classes of curves.

Vectors are independent of the choice of coordinates, hence we write

$$X = X^\mu \frac{\partial}{\partial x^\mu} = \tilde{X}^\mu \frac{\partial}{\partial y^\mu}$$

for another set of coordinates (y^μ), with X^μ and $\tilde{X}^\mu$ related via $\tilde{X}^\mu = X^\nu \frac{\partial y^\mu}{\partial x^\nu}$. It is very useful to define the pairing

$$\left\langle dx^\nu, \frac{\partial}{\partial x^\mu} \right\rangle = \frac{\partial x^\nu}{\partial x^\mu} = \delta^\nu_\mu,$$

which leads us to 1-*forms* (or covectors) $\omega = \omega_\mu dx^\mu$, also independent of choice of coordinates, and extends by linearity to the pairing between vectors (or vector fields) and 1-forms:

$$i_X \omega := \langle \omega, X \rangle = \omega_\mu X^\mu.$$

Now, under a change of local coordinates, we have

$$\omega = \omega_\mu dx^\mu = \tilde{\omega}_\nu dy^\nu \implies \tilde{\omega}_\nu = \omega_\mu \frac{\partial x^\mu}{\partial y^\nu}. \tag{2.1}$$

This can be generalized further to tensors $T^{\mu_1 \cdots \mu_q}{}_{\nu_1 \ldots \nu_r}$ of type (q, r), with q upper (contravariant) and r lower (covariant) indices.

2.2. *Tangent bundle*

Let M be an n-dimensional manifold. At each point $p \in M$, tangent vectors can be added and multiplied by real numbers, so the set of all tangent vectors forms an n-dimensional vector space, denoted $T_p M$. For a set of local coordinates (x^μ) at p, a basis of $T_p M$ is given by $\partial/\partial x^\mu$, $1 \leq \mu \leq n$. The union of all tangent spaces forms the tangent bundle

$$TM = \bigcup_{p \in M} T_p M.$$

TM is a $2n$-dimensional manifold with base space M and fibre $\mathbb{R}^n$. It is an example of an important kind of fibre bundle known as a *vector bundle*.

Similarly, at each point $p \in M$ the 1-forms form the cotangent space $T_p{}^* M$, which is the dual space to $T_p M$. The union of all cotangent spaces forms the cotangent bundle

$$T^* M = \bigcup_{p \in M} T_p^* M,$$

another $2n$-dimensional manifold which is a vector bundle with base space M and fibre $\mathbb{R}^n$. More generally, at each point p, tensors of type (q, r) are elements of the tensor product $T_pM \otimes \cdots \otimes T_pM \otimes T_p^*M \otimes \cdots \otimes T_p^*M$ with q copies of T_pM and r copies of T_p^*M, and the corresponding tensor bundle is defined similarly.

Example 2.4 (The tangent bundle TS^2). To construct the tangent bundle of S^2 we use the two stereographic projections as our charts. The coordinates $(X_1, X_2) \in U_1'$ and $(Y_1, Y_2) \in U_2'$ are related via (1.1). Given $u \in TS^2$ with $\pi(u) = p \in U_1 \cap U_2$, then the local trivializations ϕ_1 and ϕ_2 satisfy $\phi_1^{-1}(u) = (p, V_1^\mu)$ and $\phi_2^{-1}(u) = (p, V_2^\mu)$. The local trivialization is

$$t_{12} = \frac{\partial(Y_1, Y_2)}{\partial(X_1, X_2)} = \frac{1}{(X_1^2 + X_2^2)^2} \begin{pmatrix} X_2^2 - X_1^2 & -2X_1X_2 \\ -2X_1X_2 & X_1^2 - X_2^2 \end{pmatrix}.$$

The reader may check that $t_{21}(p) = t_{12}^{-1}(p)$.

Given a smooth map $f : M \to N$ between manifolds, we can define a map between tangent spaces,

$$f_* : T_pM \to T_{f(p)}N$$
$$V \mapsto f_*V,$$

which is called the *pushforward* of f. If $g \in C^\infty(N)$ then $g \circ f \in C^\infty(M)$, and the action of the vector f_*V on g is given by

$$f_*V(g) = V(g \circ f).$$

Similarly, there is a map between cotangent spaces,

$$f^* : T^*{}_{f(p)}N \to T^*{}_pM$$
$$\omega \mapsto f^*\omega$$

called the *pullback*. The pullback can be defined via the pairing between vectors and 1-forms: $\langle f^*\omega, V \rangle_M = \langle \omega, f_*V \rangle_N$.

2.3.   Sections

Definition 2.5. Let (E, M, π) be a fibre bundle. A section $s : M \to E$ is a smooth map which satisfies $\pi \circ s = id_M$. Here, $s|_p$ is an element of the fibre $F_p = \pi^{-1}(p)$. The space of sections is denoted by $\Gamma(E)$.

A local section is defined on a subset $U \subset M$, only. Not all fibre bundles admit global sections (defined on the whole of M).

Example 2.6 (Vector and tensor fields). A vector field V on a manifold M is a section of the tangent bundle: $V \in \Gamma(TM)$, associating a tangent vector to each point in M. In local coordinates $x = (x^\mu)$ this means $V = V^\mu(x)\partial/\partial x^\mu$, where the components $V^\mu(x)$ define a map from (part of) $\mathbb{R}^n \to \mathbb{R}^n$. Similarly, a differential 1-form is a section of T^*M, and a tensor field of type (q, r) is a section of the corresponding tensor bundle.

Example 2.7 (Wave functions). The wave function $\psi(\mathbf{x}, t)$ of a single particle in quantum mechanics can be thought of as a section of a complex line bundle $E = \mathbb{R}^{3,1} \times \mathbb{C}$.

Vector bundles always have at least one section, the null section s_0 with

$$\phi_i^{-1}(s_0(p)) = (p, 0)$$

in any local trivialization. A principal bundle E only admits a global section if it is trivial: $E = M \times G$. A section in a principal bundle can be used to construct the trivialization of the bundle, using the fact that there is a right action which is independent of the local trivialization: $ua = \phi_i(p, g_i a)$ for any $a \in G$.

2.4. *Associated vector bundle*

Given a principal fibre bundle P with base M, $\pi : P \to M$ with $\pi^{-1}(p) \cong G$, and a k-dimensional representation $\rho : G \to \mathrm{Aut}(V)$ of G on a vector space V, the *associated vector bundle* $E = P \times_\rho V$ is defined by identifying the points

$$(u, v) \quad \text{and} \quad (ug, \rho(g)^{-1}v) \in P \times V,$$

where $u \in P$, $g \in G$, and $v \in V$. The projection $\pi_E : E \to M$ is defined by $\pi_E(u, v) = \pi(u)$, which is well defined because

$$\pi_E(ug, \rho(g)^{-1}v) = \pi(ug) = \pi(u) = \pi_E(u, v).$$

The transition functions of E are given by $\rho(t_{ij}(p))$ where $t_{ij}(p)$ are the transition functions of P. Conversely, a vector bundle naturally induces a principal bundle associated with it.

2.5. *Metric tensor*

Manifolds can carry further structure, an important example being a *metric*. A metric g is a $(0, 2)$ tensor which, at each point $p \in M$, must satisfy the following conditions, for all $U, V \in T_p M$:

(1) $g_p(U, V) = g_p(V, U)$;
(2) $g_p(U, U) \geq 0$, with equality only when $U = 0$.

The metric g provides an inner product on each tangent space $T_p M$, and (M, g) is called a Riemannian manifold.[c] The metric gives an isomorphism between vector fields $X \in \Gamma(TM)$ and 1-forms $\eta \in \Gamma(T^*M)$ via

$$g(\cdot, X) = \eta_X.$$

In physics notation $g_{\mu\nu}$ and its inverse $g^{\mu\nu}$ lower and raise indices, and one writes

$$g = ds^2 = g_{\mu\nu} \, dx^\mu \, dx^\nu.$$

If M is a submanifold of N with metric g_N and $f : M \to N$ is the embedding map, then the induced metric g_M is

$$g_{M\mu\nu}(x) = g_{N\alpha\beta}(f(x)) \frac{\partial f^\alpha}{\partial x^\mu} \frac{\partial f^\beta}{\partial x^\nu}.$$

2.6. *Connection on the tangent bundle*

The "derivative" of a vector field $V = V^\mu \frac{\partial}{\partial x^\mu}$ with respect to x^ν is

$$\frac{\partial V^\mu}{\partial x^\nu} = \lim_{\Delta x^\nu \to 0} \frac{V^\mu(\ldots, x^\nu + \Delta x^\nu, \ldots) - V^\mu(\ldots, x^\nu, \ldots)}{\Delta x^\nu}.$$

This does not work as the first vector is defined at $x + \Delta x$ and the second at x. We need to transport the vector V^μ from x to $x + \Delta x$ "without change." This is known as *parallel transport*, and is achieved by specifying a *connection* $\Gamma^\mu{}_{\nu\lambda}$, namely the parallel transported vector $\tilde{V}^\mu$ is given by

$$\tilde{V}^\mu(x + \Delta x) = V^\mu(x) - V^\lambda(x) \Gamma^\mu{}_{\nu\lambda}(x) \Delta x^\nu.$$

The covariant derivative of V with respect to x^ν is

$$\lim_{\Delta x^\nu \to 0} \frac{V^\mu(x + \Delta x) - \tilde{V}^\mu(x + \Delta x)}{\Delta x^\nu} = \frac{\partial V^\mu}{\partial x^\nu} + V^\lambda \Gamma^\mu{}_{\nu\lambda}.$$

---

[c]Condition (2) defines a Riemannian metric. If g is non-degenerate but of indefinite signature, it is called pseudo-Riemannian; this is required for special and general relativity.

In the case where a manifold comes equipped with a (pseudo-) Riemannian metric g, it can be required that the metric g is covariantly constant, meaning that if two vectors X and Y are parallel transported along any curve, then the inner product $g(X, Y)$ remains constant. The condition

$$\nabla_V(g(X, Y)) = 0,$$

yields the *Levi-Civita connection*. The Levi-Civita connection is specified locally by the *Christoffel symbols*, written as

$$\Gamma^\kappa{}_{\mu\nu} = \tfrac{1}{2} g^{\kappa\lambda} \left(\partial_\mu g_{\nu\lambda} + \partial_\nu g_{\mu\lambda} - \partial_\lambda g_{\mu\nu} \right). \tag{2.2}$$

Example 2.8 (General relativity). The Christoffel symbols do not transform like a tensor. However, from them, one builds the Riemann curvature tensor:

$$R^\kappa{}_{\lambda\mu\nu} = \partial_\mu \Gamma^\kappa{}_{\nu\lambda} - \partial_\nu \Gamma^\kappa{}_{\mu\lambda} + \Gamma^\eta{}_{\nu\lambda} \Gamma^\kappa{}_{\mu\eta} - \Gamma^\eta{}_{\mu\lambda} \Gamma^\kappa{}_{\nu\eta}.$$

Important contractions of the curvature tensor are the *Ricci tensor Ric*,

$$Ric_{\mu\nu} = R^\lambda{}_{\mu\lambda\nu},$$

and the *scalar curvature $\mathcal{R}$*,

$$\mathcal{R} = g^{\mu\nu} Ric_{\mu\nu}.$$

Now, we have the ingredients for *Einstein's field equations of general relativity*, namely

$$Ric_{\mu\nu} - \tfrac{1}{2} g_{\mu\nu} \mathcal{R} = 8\pi G T_{\mu\nu},$$

where G is the gravitational constant and $T_{\mu\nu}$ is the energy–momentum tensor which describes the distribution of matter in spacetime.

2.7. *Yang–Mills theory and fibre bundles*

An example of Yang–Mills theory is given by the Lagrangian density

$$\mathcal{L} = \tfrac{1}{8} \mathrm{Tr} \left(\mathbf{F}_{\mu\nu} \mathbf{F}^{\mu\nu} \right) + \tfrac{1}{2} \left(D_\mu \Phi \right)^\dagger D^\mu \Phi - U(\Phi^\dagger \Phi), \tag{2.3}$$

where $U(\cdot)$ is a potential energy density,

$$D_\mu \Phi = \partial_\mu \Phi + \mathbf{A}_\mu \Phi, \quad \text{and} \quad \mathbf{F}_{\mu\nu} = \partial_\mu \mathbf{A}_\nu - \partial_\nu \mathbf{A}_\mu + [\mathbf{A}_\mu, \mathbf{A}_\nu].$$

Here Φ is a two-component complex scalar field (Higgs field), while $\mathbf{A}_\mu$ is called a gauge field and is $\mathfrak{su}(2)$-valued, i.e. $\mathbf{A}_\mu$ are anti-Hermitian 2×2 matrices; $\mathbf{F}_{\mu\nu}$ is known as the field strength (also $\mathfrak{su}(2)$-valued). This Lagrangian is Lorentz invariant: indices are raised and lowered with the Minkowski metric $g_{\mu\nu} = \mathrm{diag}(1, -1, -1, -1)$, and the Lagrangian remains the same under linear coordinate transformations which preserve this quadratic form.

The Lagrangian (2.3) is also invariant under local *gauge transformations*: Let $\mathbf{g} \in \mathrm{SU}(2)$ be a spacetime dependent gauge transformation with

$$\Phi \mapsto \mathbf{g}\Phi, \quad \text{and} \quad \mathbf{A}_\mu \mapsto \mathbf{g}\mathbf{A}_\mu\mathbf{g}^{-1} - \partial_\mu\mathbf{g}\cdot\mathbf{g}^{-1}.$$

The covariant derivative $D_\mu\Phi$ transforms as

$$D_\mu\Phi \mapsto \partial_\mu(\mathbf{g}\Phi) + \left(\mathbf{g}\mathbf{A}_\mu\mathbf{g}^{-1} - \partial_\mu\mathbf{g}\cdot\mathbf{g}^{-1}\right)\mathbf{g}\Phi = \mathbf{g}D_\mu\Phi$$

Hence $\Phi^\dagger\Phi \mapsto (\mathbf{g}\Phi)^\dagger\mathbf{g}\Phi = \Phi^\dagger\mathbf{g}^\dagger\mathbf{g}\Phi = \Phi^\dagger\Phi$, and similarly for $(D_\mu\Phi)^\dagger D^\mu\Phi$. Finally, $\mathbf{F}_{\mu\nu} \mapsto \mathbf{g}\mathbf{F}_{\mu\nu}\mathbf{g}^{-1}$, so $\mathrm{Tr}\left(\mathbf{F}_{\mu\nu}\mathbf{F}^{\mu\nu}\right)$ is also gauge invariant.

Yang–Mills theory can also be described in a more mathematical language: The gauge field $\mathbf{A}_\mu$ corresponds to a connection on a principal $\mathrm{SU}(2)$ bundle. The field strength $\mathbf{F}_{\mu\nu}$ corresponds to the curvature of the principal $\mathrm{SU}(2)$ bundle. The Higgs field Φ is a section of the associated $\mathbb{C}^2$ vector bundle. The action of $\mathbf{g} \in \mathrm{SU}(2)$ on Φ and $\mathbf{A}_\mu$ is precisely what is expected for an associated fibre bundle. Surprisingly, mathematicians and physicists derived the same result very much independently!

3. Forms and Integration

3.1. *Differential forms*

Differential r-forms are $(0, r)$ tensor fields which are completely antisymmetric, that is, they acquire a minus sign under interchange of any two indices. A basis for $\Omega^r(M)$, the set of r-forms on an n-dimensional manifold M, is given in local coordinates at each point by

$$\{dx^{\mu_1} \wedge \cdots \wedge dx^{\mu_r} \mid 1 \leq \mu_1 < \cdots < \mu_r \leq n\}.$$

Setting $\Omega^0(M) = C^\infty(M)$, for $0 \leq k, l \leq n$ the *wedge product* gives a bilinear map

$$\wedge : \Omega^k \times \Omega^l \to \Omega^{k+l},$$

satisfying associativity, i.e. $(\alpha \wedge \beta) \wedge \gamma = \alpha \wedge (\beta \wedge \gamma)$, and the property

$$\alpha \wedge \beta = (-1)^{kl} \beta \wedge \alpha.$$

The *exterior derivative* $d : \Omega^k \to \Omega^{k+1}$ can be defined as follows: Given

$$\omega = \frac{1}{k!} \omega_{\mu_1 \ldots \mu_k} \, dx^{\mu_1} \wedge \cdots \wedge dx^{\mu_k} \in \Omega^k,$$

its exterior derivative $d\omega \in \Omega^{k+1}$ is

$$d\omega = \frac{1}{k!} \left(\frac{\partial}{\partial x_\nu} \omega_{\mu_1 \ldots \mu_k} \right) dx^\nu \wedge dx^{\mu_1} \wedge \cdots \wedge dx^{\mu_k}.$$

From this definition it is an easy exercise to show that $d^2\omega = 0$ for any differential k-form ω.

Example 3.1 (Symplectic form). A symplectic form ω is a 2-form which satisfies

(1) ω is closed, i.e. $d\omega = 0$;
(2) ω is non-degenerate: at each point $p \in M$, $\omega(U, V) = 0$ for all $V \in T_pM$ implies $U = 0$.

In local coordinates, ω can be written as

$$\omega = \tfrac{1}{2} \omega_{\mu\nu} \, dx^\mu \wedge dx^\nu.$$

Like a metric g, a symplectic form also provides an isomorphism between vector fields and 1-forms via

$$\begin{aligned} \Gamma(TM) &\to \Gamma(T^*M) \\ V &\mapsto \omega(\cdot, V), \end{aligned} \tag{3.1}$$

and a manifold M equipped with a symplectic form is called a *symplectic manifold*.

3.2. *Integration on manifolds*

Recall that, under change of coordinates, 1-forms transform according to (2.1). Two charts define the same orientation provided that

$$\det \left(\frac{\partial x^\mu}{\partial y^\nu} \right) > 0. \tag{3.2}$$

A manifold is orientable if for any overlapping charts U_i and U_j there exist local coordinates x^μ for U_i and y^μ for U_j such that (3.2) holds. The invariant volume element on a (pseudo-)Riemannian manifold M is given by

$$\Omega = \sqrt{|g|}\, dx^1 \wedge \cdots \wedge dx^m \quad \text{where } |g| = |\det(g_{\mu\nu})|. \tag{3.3}$$

Given Ω and an orientation, a function $f : M \to \mathbb{R}$ can be integrated over M. For one chart,

$$\int_{U_i} f\Omega = \int_{\phi_i(U_i)} f(\phi_i^{-1}(x))\sqrt{|g(\phi_i^{-1}(x)|}\, dx^1\, dx^2 \ldots dx^m.$$

For the whole of M, one takes a *partition of unity*, which is a family of differentiable functions $\epsilon_i(p)$, $1 \le i \le k$ such that

(1) $0 \le \epsilon_i(p) \le 1$;
(2) $\epsilon_i(p) = 0$ if $p \notin U_i$;
(3) $\epsilon_1(p) + \cdots + \epsilon_k(p) = 1$ for any point $p \in M$.

Then the integral of f over the whole manifold M, covered by k charts, is

$$\int_M f\Omega = \sum_{i=1}^{k} \int_{U_i} f(p)\epsilon_i(p)\Omega.$$

A fundamental result on integration is the generalized Stokes theorem.

Theorem 3.2. *If ω is an r-form and R an $r+1$-dimensional region in M with boundary ∂R, then*

$$\int_R d\omega = \int_{\partial R} \omega.$$

Example 3.3 (Green's theorem). For $\omega = p\, dx + q\, dy$ in $\mathbb{R}^2$,

$$d\omega = (\partial_x q - \partial_y p)\, dx \wedge dy.$$

Hence, integrating around a closed curve $\mathcal{C}$ enclosing a region $R \subset \mathbb{R}^2$ gives

$$\oint_{\mathcal{C}} (p\, dx + q\, dy) = \iint_R (\partial_x q - \partial_y p)\, dx\, dy,$$

which is Green's theorem in the plane.

Example 3.4 (Stokes theorem and divergence theorem). In $\mathbb{R}^3$ with $\omega = f_1 \, dx + f_2 \, dy + f_3 \, dz$, we have

$$d\omega = (\partial_y f_3 - \partial_z f_2) dy \wedge dz + (\partial_z f_1 - \partial_x f_3) dz \wedge dx + (\partial_x f_2 - \partial_y f_1) dx \wedge dy,$$

which gives rise to the traditional Stokes theorem in vector form:

$$\oint_C \mathbf{f} \cdot d\mathbf{r} = \iint_S (\nabla \wedge \mathbf{f}) \cdot \mathbf{n} \, dS.$$

On the other hand, if $\omega = f_1 \, dy \wedge dz + f_2 \, dz \wedge dx + f_3 \, dx \wedge dy$ then

$$d\omega = (\partial_x f_1 + \partial_y f_2 + \partial_z f_3) \, dx \wedge dy \wedge dz,$$

which yields the divergence theorem:

$$\iiint_V \nabla \cdot \mathbf{f} \, dx \, dy \, dz = \iint_S \mathbf{f} \cdot \mathbf{n} \, dS.$$

When a manifold has a metric, there is another important operation on forms called the *Hodge star*. The totally antisymmetric tensor is defined by

$$\epsilon_{\mu_1 \mu_2 \ldots \mu_n} = \begin{cases} +1 & \text{if } (\mu_1 \mu_2 \ldots \mu_n) \text{ is an even permutation of } (12 \ldots n), \\ -1 & \text{if } (\mu_1 \mu_2 \ldots \mu_n) \text{ is an odd permutation of } (12 \ldots n), \\ 0 & \text{otherwise.} \end{cases}$$

The Hodge star is a linear map $* : \Omega^r(M) \to \Omega^{n-r}(M)$ which acts on a basis vector in $\Omega^r(M)$ according to

$$*(dx^{\mu_1} \wedge \cdots \wedge dx^{\mu_r}) = \frac{\sqrt{|g|}}{n!} \epsilon^{\mu_1 \ldots \mu_r}{}_{\nu_{r+1} \ldots \nu_n} dx^{\nu_{r+1}} \wedge \cdots \wedge dx^{\nu_n}.$$

The invariant volume element in (3.3) is $\Omega = *1$.

Example 3.5 (Hodge star on $\mathbb{R}^3$). In the Euclidean space $\mathbb{R}^3$,

$$*1 = dx \wedge dy \wedge dz, *dx = dy \wedge dz, *dy = dz \wedge dx, *dz = dx \wedge dy,$$

$$*dy \wedge dz = dx, *dz \wedge dx = dy, *dx \wedge dy = dz, *dx \wedge dy \wedge dz = 1.$$

Now suppose that (M, g) is Riemannian, $\dim M = n$ and ω is an r-form. Then applying the Hodge star twice gives

$$**\omega = (-1)^{r(n-r)} \omega,$$

which shows that $*$ is an isomorphism. Given the coordinate expressions

$$\omega = \frac{1}{r!} \omega_{\mu_1 \ldots \mu_r} dx^{\mu_1} \wedge \cdots \wedge dx^{\mu_r} \quad \text{and} \quad \eta = \frac{1}{r!} \eta_{\mu_1 \ldots \mu_r} dx^{\mu_1} \wedge \cdots \wedge dx^{\mu_r},$$

in terms of the volume form (3.3) we have

$$\omega \wedge *\eta = \cdots = \frac{1}{r!} \omega_{\mu_1 \dots \mu_r} \eta^{\mu_1 \dots \mu_r} \, \Omega.$$

Thus an inner product on r-forms is defined via

$$(\omega, \eta) = \int_M \omega \wedge *\eta, \tag{3.4}$$

and this product is symmetric and positive definite.

Example 3.6 (Ginzburg–Landau potential). Ginzburg–Landau vortices on $\mathbb{R}^2$ are minimals of the potential energy functional

$$V(\phi, a) = \frac{1}{2} \int_{\mathbb{R}^2} \left(da \wedge *da + \overline{d_a \phi} \wedge *d_a \phi + \frac{\lambda}{4}(1 - \bar{\phi}\phi)^2 * 1 \right), \tag{3.5}$$

where $\phi : \mathbb{R}^2 \to \mathbb{C}$ is a complex scalar field, $a = a_1 \, dx^1 + a_2 \, dx^2 \in \Omega^1(\mathbb{R}^2)$ is the gauge potential 1-form, $d_a \phi = d\phi - ia\phi$, and $*$ is the Hodge isomorphism. In usual physics notation,

$$V = \frac{1}{2} \int \left(\frac{1}{2} f^{ij} f_{ij} + \overline{D^i \phi} D_i \phi + \frac{\lambda}{4}(1 - \bar{\phi}\phi)^2 \right) dx^2,$$

where $D_i \phi = \partial_i \phi - ia_i \phi$ and $f_{12} = \partial_1 a_2 - \partial_2 a_1$.

Given the exterior derivative $d : \Omega^{r-1}(M) \to \Omega^r(M)$ we can define the adjoint exterior derivative $d^\dagger : \Omega^r(M) \to \Omega^{r-1}(M)$ via

$$d^\dagger = (-1)^{nr+n+1} * d *.$$

If (M, g) is compact, orientable and without boundary, and $\alpha \in \Omega^r(M)$, $\beta \in \Omega^{r-1}(M)$ then

$$(d\beta, \alpha) = (\beta, d^\dagger \alpha),$$

so $d^\dagger$ is the adjoint with respect to the inner product (3.4). The Laplacian $\Delta : \Omega^r(M) \to \Omega^r(M)$ is defined by

$$\Delta = (d + d^\dagger)^2 = dd^\dagger + d^\dagger d.$$

For a function $f : M \to \mathbb{R}$, it is a useful exercise to show that in coordinates the Laplacian is given by

$$\Delta f = -\frac{1}{\sqrt{|g|}} \, \partial_\nu(\sqrt{|g|} g^{\mu\nu} \partial_\mu f).$$

An r-form ω_r is called harmonic if $\Delta\omega_r = 0$, and the set of harmonic r-forms is denoted by $\mathrm{Harm}^r(M)$.

Theorem 3.7 (Hodge decomposition).

$$\Omega^r(M) = d\Omega^{r-1}(M) \oplus d^\dagger\Omega^{r+1} \oplus \mathrm{Harm}^r(M),$$

that is, any $\omega_r \in \Omega^r(M)$ can be decomposed as

$$\omega_r = d\alpha_{r-1} + d^\dagger\beta_{r+1} + \gamma_r$$

with $\Delta\gamma_r = 0$.

In fact, $\mathrm{Harm}^r(M) \cong H^r(M)$, where $H^r(M)$ is the de Rham cohomology group (closed r-forms modulo exact r-forms).

Example 3.8 (Maxwell's equations). The four Maxwell equations for the electric field $\mathbf{E}$ and magnetic field $\mathbf{B}$ can be written in vector form as

$$\nabla \cdot \mathbf{E} = \rho, \quad \nabla \wedge \mathbf{B} - \frac{\partial \mathbf{E}}{\partial t} = \mathbf{j}, \quad \nabla \cdot \mathbf{B} = 0, \quad \text{and} \quad \nabla \wedge \mathbf{E} + \frac{\partial \mathbf{B}}{\partial t} = 0,$$

where, in terms of the electromagnetic potential $(A_0, \mathbf{A})$,

$$\mathbf{E} = -\nabla A_0 - \frac{\partial \mathbf{A}}{\partial t} \quad \text{and} \quad \mathbf{B} = \nabla \wedge \mathbf{A}.$$

In differential geometry notation, the potential is encoded into a 1-form A, and the electromagnetic field tensor is the 2-form $F = dA$. With a 1-form current j, the Maxwell equations are just

$$d^\dagger F = j \quad \text{and} \quad dF = 0.$$

3.3. *Complex manifolds*

If $z = x + iy$ and $f = u + iv$ then $f(x, y)$ is *holomorphic* in z provided the Cauchy–Riemann equations are satisfied:

$$\frac{\partial u}{\partial x} = \frac{\partial v}{\partial y}, \quad \frac{\partial u}{\partial y} = -\frac{\partial v}{\partial x}.$$

A complex manifold is a manifold such that the crossover maps ψ_{ij} are all holomorphic. Examples of complex manifolds are $\mathbb{C}^n$, S^2, T^2, $\mathbb{C}P^n$, and $S^{2n+1} \times S^{2m+1}$.

An *almost complex structure* is a $(1,1)$ tensor field J which acts linearly on T_pM in terms of real coordinates as

$$J_p \frac{\partial}{\partial x^\mu} = \frac{\partial}{\partial y^\mu}, \quad J_p \frac{\partial}{\partial y^\mu} = -\frac{\partial}{\partial x^\mu} \tag{3.6}$$

with $J_p{}^2 = -\mathrm{id}_{T_pM}$.[d] In terms of complex coordinate vectors we have

$$J_p \frac{\partial}{\partial z^\mu} = \mathrm{i}\frac{\partial}{\partial z^\mu}, \quad J_p \frac{\partial}{\partial \bar{z}^\mu} = -\mathrm{i}\frac{\partial}{\partial \bar{z}^\mu}.$$

(multiplication by $\mathrm{i} = \sqrt{-1}$).

A *Hermitian metric* is a Riemannian metric which satisfies

$$g_p(J_pX, J_pY) = g_p(X,Y),$$

i.e. g is compatible with J_p. The vector J_pX is orthogonal to X:

$$g_p(J_pX, X) = g_p(J_p^2 X, J_pX) = -g_p(J_pX, X) = 0.$$

For a Hermitian metric $g_{\mu\nu} = 0$ and $g_{\bar{\mu}\bar{\nu}} = 0$, e.g.

$$g_{\mu\nu} = g\left(\frac{\partial}{\partial z^\mu}, \frac{\partial}{\partial z^\nu}\right) = g\left(J_p\frac{\partial}{\partial z^\mu}, J_p\frac{\partial}{\partial z^\nu}\right) = g\left(\mathrm{i}\frac{\partial}{\partial z^\mu}, \mathrm{i}\frac{\partial}{\partial z^\nu}\right) = -g_{\mu\nu}.$$

Given the metric and an (almost) complex structure, define ω via

$$\omega_p(X,Y) = g_p(J_pX, Y), \quad X,Y \in T_pM.$$

Then ω is an antisymmetric tensor field, and invariant under J_p :

$$\omega(X,Y) = -\omega(Y,X), \quad \omega(J_pX, J_pY) = \omega(X,Y).$$

Moreover, ω is a real form and can be written as

$$\omega = -\mathrm{i}g_{\mu\bar{\nu}}dz^\mu \wedge d\bar{z}^\nu.$$

Also, $\omega \wedge \cdots \wedge \omega$ ($\dim_{\mathbb{C}} M$ times) provides a volume form for M. If $d\omega = 0$ then g is called a *Kähler metric*. For a Kähler manifold, the metric g is related to the antisymmetric Kähler form ω which can be interpreted as a symplectic 2-form.

---

[d]Compare this with the matrix appearing in the right-hand side of (4.9) below.

Remark 3.9. As we shall see later, topological solitons of Bogomolny type usually have a "moduli space" of static solutions which is a smooth manifold with a natural Kähler metric given by the kinetic energy.

4. Geometry in Classical Mechanics

The language of differential geometry is extremely useful for formulating classical mechanics. There are two main approaches: Lagrangian mechanics and Hamiltonian mechanics; and these two approaches lead to two different geometrical settings: symplectic geometry and Poisson geometry.

4.1. *Lagrangian mechanics*

The traditional formulation of Lagrangian mechanics involves two manifolds, namely the configuration space Q together with its tangent bundle TQ. Suppose that Q has dimension n, and local coordinates $(\mathbf{q}, \mathbf{v}) = (q^1, q^2, \ldots, q^n, v^1, v^2, \ldots, v^n)$ are given on TQ. Then the motion of a system is specified by the Lagrangian, which is a function

$$L: \quad TQ \to \mathbb{R},$$

so locally $L = L(\mathbf{q}, \mathbf{v})$. If the velocity $\mathbf{v} = \dot{\mathbf{q}}$ is the tangent vector to a path γ in Q parametrized by time t (with dot denoting d/dt), then we can write $L = L(\mathbf{q}, \dot{\mathbf{q}})$ and consider the function L along the path γ. The action $\mathcal{S}$ associated with L is given by integrating it between two fixed, arbitrary times $t_0 < t_1$ to obtain

$$\mathcal{S} = \int_{t_0}^{t_1} L(\mathbf{q}(t), \dot{\mathbf{q}}(t)) \, dt,$$

and the Principle of Least Action selects those paths which are stationary under variation of the action, i.e. $\delta\mathcal{S} = 0$. By a standard result of the calculus of variations [9], this yields the Euler–Lagrange equations

$$\frac{d}{dt}\left(\frac{\partial L}{\partial \dot{q}^j}\right) - \frac{\partial L}{\partial q^j} = 0, \quad j = 1, \ldots, n. \tag{4.1}$$

Example 4.1 (Harmonic oscillator). With $Q = \mathbb{R}$, a harmonic oscillator of mass m and frequency ω is specified by the Lagrangian $L = \frac{1}{2}m\dot{q}^2 - \frac{1}{2}m\omega^2 q^2$.

Example 4.2 (Simple pendulum). For $Q = S^1$ with angular coordinate $q \in (-\pi, \pi)$, the Lagrangian for a simple pendulum of mass m and length ℓ experiencing gravitational acceleration g is $L = \frac{1}{2}m\ell^2\dot{q}^2 - mg\ell(1 - \cos q)$.

Example 4.3 (Newton's second law). The preceding two examples are particular cases of a Lagrangian of the form $L = \frac{1}{2}\sum_j m_j(\dot{q}^j)^2 - V(\mathbf{q})$ (kinetic energy minus potential energy), with the Euler–Lagrange equations (4.1) being

$$m_j\ddot{q}^j = -\frac{\partial V}{\partial q^j}, \quad j = 1, \ldots, n,$$

corresponding to Newton's second law with conservative forces described by the potential V.

Example 4.4 (Geodesic flow). Given a Riemannian manifold Q with metric g given in local coordinates as $ds^2 = g_{jk}\,dq^j\,dq^k$, consider $\mathcal{S} = \int_{t_0}^{t_1} L\,dt$ with the purely kinetic Lagrangian

$$L = \tfrac{1}{2}g(\dot{\mathbf{q}}, \dot{\mathbf{q}}) = \tfrac{1}{2}g_{jk}(\mathbf{q})\dot{q}^j\dot{q}^k \tag{4.2}$$

(the summation convention is assumed throughout). Then (4.1) produces

$$\ddot{q}^j + \Gamma^j_{kl}\dot{q}^k\dot{q}^l = 0, \tag{4.3}$$

where Γ^j_{kl} are the Christoffel symbols (2.2). These geodesic equations can also be obtained as the paths of minimum length [5], by taking the alternative action $\int_{t_0}^{t_1} ds = \int_{t_0}^{t_1} \sqrt{g(\dot{\mathbf{q}}, \dot{\mathbf{q}})}\,dt$. The latter interpretation no longer applies in the pseudo-Riemannian setting, since g is not positive definite, e.g. in the case of general relativity with $n = 4$, where g has the Lorentzian signature $(+, -, -, -)$, and equations (4.3) describe a free particle moving through spacetime [15].

4.2. *Hamiltonian mechanics: canonical case*

The Euler–Lagrange equations (4.1) are ordinary differential equations of second order for the generalized coordinates $\mathbf{q}$. An alternative formulation of mechanics, due to Hamilton, is framed in terms of differential equations of first order. The canonical version of Hamiltonian mechanics can be derived from the Lagrangian setting by performing a Legendre transformation, defining the generalized momenta $\mathbf{p} = (p_1, p_2, \ldots, p_n)$ according to

$$p_j = \frac{\partial L}{\partial v^j}, \quad j = 1, \ldots, n, \tag{4.4}$$

and introducing

$$H(\mathbf{q}, \mathbf{p}) = \langle \mathbf{p}, \mathbf{v} \rangle - L(\mathbf{q}, \mathbf{v}), \tag{4.5}$$

with $\langle, \rangle$ denoting the standard scalar product (so in components $H = p_j v^j - L$). In general, the Hamiltonian H can only be found as a function of $\mathbf{p}$ (and $\mathbf{q}$) if (4.4) can be inverted to find $\mathbf{v} = \mathbf{v}(\mathbf{q}, \mathbf{p})$; so the Hessian matrix $\left(\frac{\partial^2 L}{\partial v^j \partial v^k} \right)$ must be non-singular, in order to apply the implicit function theorem.

Theorem 4.5. *Given momenta p_j and velocities $v^j = \dot{q}^j$ related by (4.4), and the corresponding Legendre transformation (4.5) between functions L and H, the Euler–Lagrange equations (4.1) hold for $\mathbf{q}$ if and only if Hamilton's canonical equations*

$$\dot{q}^j = \frac{\partial H}{\partial p_j}, \quad \dot{p}_j = -\frac{\partial H}{\partial q_j} \tag{4.6}$$

are satisfied for $j = 1, \ldots, n$.

Proof. From (4.5) we have on the one hand

$$dH = v^j \, dp_j - \frac{\partial L}{\partial q^j} \, dq^j + \left(p_j - \frac{\partial L}{\partial v^j} \right) dv^j,$$

where the last term in brackets vanishes by (4.4), while on the other hand

$$dH = \frac{\partial H}{\partial q^j} \, dq^j + \frac{\partial H}{\partial p^j} \, dp^j.$$

Comparing these two expressions yields

$$\frac{\partial H}{\partial q^j} = -\frac{\partial L}{\partial q^j}, \quad v^j = \frac{\partial H}{\partial p^j},$$

and with $v^j = \dot{q}^j$ the result follows. □

Example 4.6 (Natural Hamiltonian). Applying the Legendre transformation to the Lagrangian in Example 4.3, we see that the Hamiltonian is

$$H = \sum_j \frac{(p_j)^2}{2m_j} + V(\mathbf{q}),$$

where the momenta are $p_j = m_j \dot{q}^j$ (mass times velocity) for $j = 1, \ldots, n$. This is an example of a natural Hamiltonian, given as a sum of kinetic and potential energy, with the kinetic term being quadratic in momenta.

Example 4.7 (Geodesic Hamiltonian). The most general kinetic term, quadratic in momenta, arises from Example 4.4. In that case, the Legendre transformation produces the Hamiltonian for geodesic flow, which is

$$H = \tfrac{1}{2} g^{jk}(\mathbf{q}) p_j p_k, \tag{4.7}$$

where g^{jk} are the components of the co-metric tensor ($g^{jk} g_{k\ell} = \delta^j_\ell$).

Geometrically, the formula (4.4) defines a map $TQ \to T^*Q$, which sends the vector $v^j \partial/\partial q^j$ to the Poincaré 1-form $\alpha = p_j dq^j$. This yields the symplectic form

$$\omega = d\alpha = dp_j \wedge dq^j, \tag{4.8}$$

which endows the phase space $M = T^*Q$ with the natural structure of a symplectic manifold. Any symplectic manifold is of even dimension, $2n$ say, and the Darboux theorem says that in the neighbourhood of any point there exist coordinates $(q^1, \ldots, q^n, p_1, \ldots, p_n)$ (called canonical coordinates and momenta, or Darboux coordinates) such that the symplectic form is given by the expression (4.8).

By contraction, the symplectic form defines the isomorphism (3.1) taking vector fields to one-forms. Thus to any function $F \in C^\infty(M)$ we can associate a Hamiltonian vector field $V_F \in \Gamma(TM)$, defined by $\omega(\cdot, V_F) = dF$. Equations (4.6) give the flow of the Hamiltonian vector field V_H, which can be written with vector notation as

$$\begin{pmatrix} \dot{\mathbf{q}} \\ \dot{\mathbf{p}} \end{pmatrix} = \begin{pmatrix} 0 & 1 \\ -1 & 0 \end{pmatrix} \begin{pmatrix} \dfrac{\partial H}{\partial \mathbf{q}} \\ \dfrac{\partial H}{\partial \mathbf{p}} \end{pmatrix}. \tag{4.9}$$

The symplectic form is preserved under Hamiltonian flow, in the sense that $\varphi_t^* \omega = \omega$ where $\varphi_t : M \to M$ is the time t flow map generated by V_H. It follows that for each k, $1 \leq k \leq n$, the $2k$-form $\omega \wedge \omega \wedge \cdots \wedge \omega$ (k times) is preserved by the flow; the particular case $k = n$ (preservation of the phase space volume form) is known as Liouville's theorem, and is a key result in statistical mechanics [16].

The inverse of the map (3.1) defines a bivector field J (a contravariant antisymmetric 2-tensor), called the Poisson tensor, which maps 1-forms to vector fields, so in particular $V_H = J(\cdot, dH)$, and in terms of Darboux coordinates the components of J are given by the matrix appearing on the right-hand side of (4.9). The Poisson tensor can be used to define the

Poisson bracket $\{,\}$ between functions, which is a bilinear skew-symmetric bracket given by

$$\{F, G\} = J(dF, dG) \qquad (4.10)$$

for $F, G \in C^\infty(M)$. This yields the canonical Poisson bracket relations between the Darboux coordinates, namely

$$\{q^j, p_k\} = \delta^j_k, \quad \{q^j, q^k\} = 0 = \{p_j, p_k\}. \qquad (4.11)$$

After quantization, the latter relations become the canonical commutation relations discovered by Born [6].[e]

The problem of explicitly integrating differential equations is a difficult one, and in general is a hopeless task. However, for a Hamiltonian system, Poisson brackets provide an algebraic way of determining first integrals (constants of motion) due to the following result.

Proposition 4.8. *A function F is a first integral of the Hamiltonian flow generated by H if and only if it is in involution with H, i.e. $\{F, H\} = 0$.*

Proof. For any function F the time evolution is given by

$$\frac{dF}{dt} = i_{V_H} dF = J(dF, dH) = \{F, H\}. \qquad (4.12)$$

Hence $\frac{dF}{dt} = 0$ if and only if the Poisson bracket of F with H vanishes. $\square$

From the skew-symmetry of the bracket, it follows that H itself is a constant of motion for the Hamiltonian flow that it generates. Thus the solution of a Hamiltonian system is restricted to a level set $H =$ constant, of codimension one in phase space. If there are additional first integrals then the motion is further restricted, and it may even be possible to integrate it completely.

Definition 4.9 (Complete integrability). A Hamiltonian system on a symplectic manifold M of dimension $2n$ is said to be completely integrable if it admits n functions $H = H_1, H_2, \ldots, H_n$ which are functionally independent $(dH_1 \wedge dH_2 \wedge \cdots \wedge dH_n \neq 0)$ and in involution with respect to the Poisson bracket on M, i.e. $\{H_j, H_k\} = 0$ for all j, k.

---

[e]In quantum mechanics, each pair of classical position and momentum variables q, p is replaced by a pair of operators Q, P on a Hilbert space of states, satisfying $[Q, P] = i\hbar 1$ (see e.g. [12, 21]).

The above definition is also referred to as integrability in the sense of Liouville, or Liouville integrability, due to another result known as Liouville's theorem, with a more contemporary proof and extension due to Arnold [1].

Theorem 4.10 (Liouville–Arnold). *If a Hamiltonian system satisfies the conditions of Definition 4.9, then Hamilton's equations for H can be solved by quadratures. Furthermore, each compact level set of the n functions $H_1, H_2, \ldots, H_n$ is diffeomorphic to an n-dimensional torus T^n.*

A key ingredient of the proof is the construction of action-angle coordinates in the neighbourhood of each level set: these are action variables I_j (functions of $H_1, H_2, \ldots, H_n$ only) and canonically conjugate angles θ_j (coordinates on a torus T^n). In these coordinates, the Hamiltonian is a function of the action variables only, and Hamilton's equations become

$$\dot{\theta}_j = \frac{\partial H}{\partial I_j}, \quad \dot{I}_j = 0, \quad j = 1, \ldots, n,$$

with the solution giving straight line motion on the torus:

$$\theta_j(t) = \frac{\partial H}{\partial I_j} t + \theta_j(0), \quad I_j = \text{constant}.$$

So for a completely integrable system, the motion is quasiperiodic on each compact level set of the first integrals.

Example 4.11 (Kepler problem). The Hamiltonian for a body of mass m moving in three dimensions in an attractive central force obeying the inverse square law is

$$H = \frac{|\mathbf{p}|^2}{2m} - \frac{\kappa}{|\mathbf{q}|}, \quad \kappa > 0,$$

where $(\mathbf{q}, \mathbf{p}) \in T^*\mathbb{R}^3 \simeq \mathbb{R}^3 \times \mathbb{R}^3$. Defining the angular momentum vector

$$\mathbf{L} = \mathbf{q} \times \mathbf{p}$$

with components L_j, it can be verified that the canonical bracket (4.11) for the positions and momenta leads to the relations

$$\{L_j, L_k\} = \epsilon_{jk\ell} L_\ell, \tag{4.13}$$

from which it can be verified that

$$\{H, |\mathbf{L}|^2\} = \{H, L_3\} = \{|\mathbf{L}|^2, L_3\} = 0,$$

so this is a completely integrable system.

The existence of first integrals is often associated with symmetries of a system; this connection can be made precise using Noether's theorem (traditionally in the Lagrangian setting). For the Kepler problem, the conservation of angular momentum is a consequence of rotation invariance, described by the Lie group SO(3), and the same is true if the potential $-\kappa/|\mathbf{q}|$ is replaced by any rotation-invariant function $V(|\mathbf{q}|)$. However, the inverse square law is special: it has an extra hidden symmetry [11], leading to an additional conserved vector, namely the Laplace–Runge–Lenz vector $\mathbf{p} \times \mathbf{L} - m\kappa\mathbf{q}/|\mathbf{q}|$.

4.3. *Hamiltonian mechanics: General case*

Rather than starting with a Lagrangian and proceeding via a Legendre transformation, there is a more general formulation of Hamiltonian mechanics which takes the Poisson bracket as the starting point. The development of this point of view was actually inspired by the infinite-dimensional case (Hamiltonian partial differential equations), which will be described below in due course.

Definition 4.12 (Poisson bracket). Given an algebra $\mathcal{F}$ over $\mathbb{R}$, the map

$$\{,\} \colon \mathcal{F} \times \mathcal{F} \to \mathcal{F}$$

is called a Poisson bracket if the following properties hold $\forall F, G, H \in \mathcal{F}$:

- **Skew-symmetry:** $\{F, G\} = -\{G, F\}$;
- **Bilinearity:** $\{\lambda F + \mu G, H\} = \lambda\{F, H\} + \mu\{G, H\} \; \forall \lambda, \mu \in \mathbb{R}$;
- **Jacobi identity:** $\{F, \{G, H\}\} + \{G, \{H, F\}\} + \{H, \{F, G\}\} = 0$;
- **Derivation (Leibniz rule):** $\{F, GH\} = \{F, G\}H + G\{F, H\}$.

Observe that, as stated, the second property of the above is just linearity in the first argument, but together with the first property this implies bilinearity of the bracket (i.e. linearity in both arguments), and the first three properties endow $\mathcal{F}$ with the structure of a Lie algebra. A manifold M equipped with a Poisson bracket on the functions $\mathcal{F} = C^\infty(M)$ is called a *Poisson manifold*. On a Poisson manifold M, a Hamiltonian vector field V_F is associated to each function by taking $V_F = \{\cdot, F\}$. If M has dimension d, then in local coordinates $\mathbf{x} = (x^1, \ldots, x^d)$ the equations of motion for the Hamiltonian H are

$$\dot{\mathbf{x}} = J(\cdot, dH), \tag{4.14}$$

where the Poisson tensor J is a bivector field given locally by

$$J = \frac{1}{2} J^{jk} \frac{\partial}{\partial x^j} \wedge \frac{\partial}{\partial x^k}, \quad J^{jk} = \{x^j, x^k\}.$$

Every symplectic manifold is also a Poisson manifold, with Poisson tensor defined by the inverse of (3.1), but otherwise the dimension d need not be even.

Example 4.13 (Lie–Poisson bracket for $\mathfrak{so}(3)$). Take $M = \mathbb{R}^3$ with coordinates $\mathbf{L} = (L_1, L_2, L_3)$ and bracket (4.13). The Poisson tensor has rank 2 at all points of $\mathbb{R}^3$ except the origin, where it vanishes. This is an example of a Lie–Poisson bracket: given any Lie algebra $\mathfrak{g}$ with basis $(X_j)_{j=1}^d$ satisfying $[X_j, X_k] = c_{jk\ell} X_\ell$ for structure constants $c_{jk\ell}$, a linear Poisson bracket on the dual space $\mathfrak{g}^*$ with coordinates (x_j) is defined by $\{x_j, x_k\} = c_{jk\ell} x_\ell$.

In general, a Poisson manifold is foliated by symplectic leaves: define an equivalence relation on the points of M by saying that $x \sim y$ if x and y are connected by piecewise smooth curves, each component of which is an integral curve of a Hamiltonian vector field; then each equivalence class is an immersed submanifold $N \subset M$, a symplectic manifold whose dimension is the rank of the Poisson tensor at any point of N [23]. For instance, in the preceding example, the symplectic leaves consist of two-dimensional spheres together with the origin (dimension zero); in this case they correspond to coadjoint orbits: the orbits of $G = \mathrm{SO}(3)$ acting on $\mathfrak{g}^* = \mathfrak{so}(3)^* \simeq \mathbb{R}^3$.

Another feature of Poisson manifolds, not arising in the symplectic setting, is that there can be non-constant functions whose differential is in the kernel of J.

Definition 4.14 (Casimir function). A function $C \in C^\infty(M)$ on a Poisson manifold M is called a Casimir function if $\{C, F\} = 0 \; \forall F \in C^\infty(M)$.

Example 4.15. In Example 4.13, the function

$$|\mathbf{L}|^2 = (L_1)^2 + (L_2)^2 + (L_3)^2 \tag{4.15}$$

is a Casimir for the $\mathfrak{so}(3)$ bracket (4.13). The level sets of this function coincide with the orbits of $\mathrm{SO}(3)$ acting on $\mathbb{R}^3$.

From Proposition 4.8 it is clear that Casimir functions provide first integrals for a Hamiltonian system. However, the algebra of Casimir functions can be very complicated, and it is not clear what should be the correct generalization of Definition 4.9 in the case of Poisson manifolds. Nevertheless, we can adopt the following definition from [22], which is sufficient to describe integrability in many situations, particularly in an algebraic context.

Definition 4.16 (Complete integrability (Poisson case)). Suppose that the Poisson tensor is of constant rank $2n$ on a dense open subset of a Poisson manifold M of dimension d, and that the algebra of Casimir functions is maximal, i.e. it contains $d - 2n$ independent functions. A Hamiltonian system on M is said to be completely integrable if it admits $d - n$ independent functions (including the Hamiltonian H) which are in involution.

Example 4.17 (Euler top). For a rigid body rotating freely about a fixed point, the angular momentum is a point in the phase space $M = \mathbb{R}^3$ with the Poisson bracket (4.13), and the Hamiltonian is

$$H = -\frac{1}{2}\left(\frac{(L_1)^2}{I_1} + \frac{(L_2)^2}{I_2} + \frac{(L_3)^2}{I_3}\right),$$

where $I = \mathrm{diag}(I_1, I_2, I_3)$ is the diagonalized inertia tensor. Hamilton's equations $\dot{L}_j = \{L_j, H\}$ for $j = 1, 2, 3$ can be written in vector form as

$$\frac{d\mathbf{L}}{dt} = \mathbf{L} \times \boldsymbol{\omega}, \tag{4.16}$$

where $\boldsymbol{\omega} = I^{-1}\mathbf{L}$ is the angular momentum. The Casimir (4.15) and H are two independent functions in involution, and the other conditions of Definition (4.16) are satisfied, so the system (4.16) is completely integrable.

Remark 4.18. The full description of the rigid body involves a rotation $R(t) \in \mathrm{SO}(3)$ which satisfies a second-order equation, corresponding to geodesic motion on the group $\mathrm{SO}(3)$ with respect to a suitable metric [14].

In the theory of integrable systems, and especially in infinite dimensions, there are numerous examples of systems with more than one Hamiltonian structure.

Definition 4.19 (Bi-Hamiltonian system). Two Poisson brackets $\{,\}_{1,2}$ are said to be compatible if any linear combination

$$\lambda_1 \{,\}_1 + \lambda_2 \{,\}_2$$

is also a Poisson bracket. A system is said to be bi-Hamiltonian if the flow
can be written as a Hamiltonian vector field with respect to two independent
Poisson brackets that are compatible with each other, i.e.

$$\{\cdot, H_1\}_1 = \{\cdot, H_2\}_2$$

for two different Hamiltonian functions $H_{1,2}$.

Note that, for two Poisson brackets to be compatible, it is enough to
check that their sum satisfies the Jacobi identity. It turns out that for the
Euler top there is another Poisson bracket that is compatible with (4.13).

Example 4.20 (Bi-Hamiltonian structure for Euler top). Denote
the bracket (4.13) by $\{,\}_1$, set $H_1 = H$, and define

$$\{L_1, L_2\}_2 = \frac{L_3}{I_3}, \quad \{L_2, L_3\}_2 = \frac{L_1}{I_1}, \quad \{L_3, L_1\}_2 = \frac{L_2}{I_2}.$$

Then $\{,\}_{1,2}$ are compatible, and equations (4.16) can be written as

$$\dot{L}_j = \{L_j, H_2\}_2, \quad j = 1, 2, 3, \text{ with } H_2 = \frac{1}{2}|\mathbf{L}|^2,$$

so the Euler top is bi-Hamiltonian.

Remark 4.21. One way to verify the preceding result is by observing that,
under the flow (4.16), any function F evolves according to

$$\frac{dF}{dt} = \det \frac{\partial(F, H_1, H_2)}{\partial(L_1, L_2, L_3)}.$$

In the above, the Jacobian determinant on the right-hand-side defines the
Nambu bracket of three functions on $\mathbb{R}^3$ [19], denoted $\{F, H_1, H_2\}$.

5.   Classical Field Theory

There are two obvious ways to generalize the variational approach to classical mechanics: firstly, one can consider Lagrangians with higher order
derivatives; and secondly, one can take derivatives with respect to additional
independent variables (space as well as time). The canonical Hamiltonian
framework for Lagrangians with higher derivatives was derived by Ostrogradsky (see [3] for applications), but higher order Lagrangians have several
features which make them undesirable for a physical theory [24]. In this
section we describe first-order Lagrangian densities for field theories and
the canonical Hamiltonian formalism, as well as more general Hamiltonian
formulations of evolutionary partial differential equations (PDEs).

5.1. *Lagrangians for scalar fields*

For a single scalar field ϕ in Minkowski spacetime M of dimension $n = d+1$, the action takes the form

$$S = \int_M \mathcal{L} \, d^{d+1}x, \tag{5.1}$$

where the Lagrangian density is

$$\mathcal{L} = \tfrac{1}{2}\partial^\mu \phi \partial_\mu \phi - \mathcal{V}(\phi), \tag{5.2}$$

with the Lorentz metric $g = \mathrm{diag}(1, -1, \ldots, -1)$, and

$$\partial_\mu \phi = \frac{\partial \phi}{\partial x^\mu}, \quad \partial^\mu \phi = g^{\mu\nu}\partial_\nu \phi;$$

and all indices are lowered and raised with $g_{\mu\nu}$ and $g^{\mu\nu}$, the components of g and its inverse (co-metric). Typically $n = 4$, and the coordinate indices are labelled from $\mu = 0$, so $(x^\mu) = (x^0, x^1, x^2, x^3) = (t, \mathbf{x})$ (setting the speed of light $c = 1$), and the Lagrangian is Lorentz invariant. With different coordinates, or in a curved spacetime, the appropriate metric should be used instead, and the volume element $d^n x = d^{d+1}x$ in (5.1) should be replaced by the invariant volume form (3.3). Different field theories result from the choice of function $\mathcal{V}$, which specifies the potential energy of the field.

Taking the action (5.1) with a first-order Lagrangian density, and applying the Principle of Least Action, $\delta S = 0$, with vanishing boundary conditions at infinity yields the Euler–Lagrange equations

$$\frac{\partial}{\partial x^\mu}\left(\frac{\partial \mathcal{L}}{\partial(\partial_\mu \phi)}\right) - \frac{\partial \mathcal{L}}{\partial \phi} = 0. \tag{5.3}$$

For $\mathcal{L}$ given by (5.2), this becomes

$$\partial^\mu \partial_\mu \phi + \mathcal{V}'(\phi) = 0. \tag{5.4}$$

Example 5.1 (Klein–Gordon field). This corresponds to the choice $\mathcal{V} = \tfrac{1}{2}m^2\phi^2$, where m is the mass. This is called a free field theory: the resulting field equation (see (5.4) below) is linear.

Example 5.2 (ϕ^4 theory). This is a nonlinear theory with the discrete symmetry $\phi \to -\phi$, which arises by setting $\mathcal{V} = \tfrac{1}{2}m^2\phi^2 - \tfrac{1}{4}\lambda\phi^4$ (with λ being a coupling constant).

Example 5.3 (Sine-Gordon theory). The choice $\mathcal{V} = m^2(1 - \cos\phi)$ results in an integrable field theory in dimension $n = 2$, both at the classical and quantum level. It has exact multiple soliton solutions called *kinks*.

For the case of a first-order Lagrangian with fields taking values in a target space of dimension > 1 (e.g. sigma models, where ϕ takes values in a Lie group, or Yang–Mills–Higgs theories like (2.3), with gauge fields), each component of each field satisfies an equation of the form (5.3).

5.2. *Hamiltonian field theory: canonical case*

In order to formulate a field theory as a Hamiltonian system, it is necessary to separate out space and time, which breaks Lorentz covariance. After rewriting the action (5.1) as

$$ S = \int L \, dt, \quad L = \int \mathcal{L} \, d^d x, $$

the total energy (Hamiltonian) is defined by the Legendre transformation

$$ H = \int_{\mathbb{R}^d} \pi \, \partial_t \phi \, d^d x - L, \quad \pi = \frac{\partial \mathcal{L}}{\partial(\partial_t \phi)}, $$

where π is called the momentum density. For the Lagrangian density (5.2) the Hamiltonian becomes

$$ H = \int_{\mathbb{R}^d} \left(\frac{1}{2}\pi^2 + \frac{1}{2}|\nabla\phi|^2 + \mathcal{V}(\phi) \right) d^d x. $$

The canonical form of Hamilton's equations is

$$ \begin{pmatrix} \partial_t \phi \\ \partial_t \pi \end{pmatrix} = \begin{pmatrix} 0 & 1 \\ -1 & 0 \end{pmatrix} \begin{pmatrix} \dfrac{\delta H}{\delta \phi} \\ \dfrac{\delta H}{\delta \pi} \end{pmatrix} \tag{5.5} $$

(cf. equation (4.9) in finite dimensions), where the *Fréchet derivative* of a functional H with respect to a field u is defined by

$$ \left\langle \frac{\delta H}{\delta u}, v \right\rangle = \frac{d}{d\epsilon} H[u + \epsilon v]|_{\epsilon=0}, $$

with $\langle , \rangle$ denoting the L^2 pairing on $\mathbb{R}^d$: $\langle f, g \rangle = \int_{\mathbb{R}^d} fg \, d^d x$.

5.3. *Generalized Hamiltonian structures for PDEs*

In the $1+1$-dimensional case ($d = 1$), we now consider PDEs involving higher spatial derivatives, which can be written as Hamiltonian systems (first order in time).[f] For a single field u, Hamilton's equations take the form

$$u_t = \mathcal{J}\frac{\delta H}{\delta u}, \tag{5.6}$$

where $\mathcal{J}$ is a skew-symmetric operator defining a Poisson bracket between pairs of functionals F, G according to

$$\{F, G\} = \left\langle \frac{\delta F}{\delta u}, \mathcal{J}\frac{\delta G}{\delta u} \right\rangle = \int \frac{\delta F}{\delta u}\, \mathcal{J}\, \frac{\delta G}{\delta u}\, dx.$$

(Integrals are taken over the whole of $\mathbb{R}$, and it is assumed that all fields and their derivatives vanish at infinity, so $\int F_x\, dx = 0$ for any quantity F.)

Example 5.4 (First Hamiltonian structure for KdV). The Korteweg–de Vries (KdV) equation, describing long waves moving on a shallow canal, is

$$u_t = u_{3x} + 6uu_x. \tag{5.7}$$

Letting D_x denote total derivative with respect to x, KdV can be written in Hamiltonian form by taking $\mathcal{J} = \mathcal{J}_1$ and $H = H_1$, where

$$\mathcal{J}_1 = D_x, \quad H_1 = \int \left(-\frac{1}{2}u_x^2 + u^3 \right) dx. \tag{5.8}$$

For PDEs, the notion of integrability can be defined by the existence of an infinite set of commuting symmetries, and in the Hamiltonian setting this is linked to infinitely many conservation laws, which can be derived recursively from a bi-Hamiltonian structure.

Example 5.5 (Second Hamiltonian structure for KdV). Let

$$\mathcal{J}_2 = D_x^3 + 4uD_x + 2u_x, \quad H_2 = \int \frac{1}{2}u^2\, dx. \tag{5.9}$$

---

[f] For convenience, here we use subscripts to denote partial derivatives, so $u_t = \frac{\partial u}{\partial t}$, $u_{xx} = u_{2x} = \frac{\partial^2 u}{\partial x^2}$, $u_{xxx} = u_{3x} = \frac{\partial^3 u}{\partial x^3}$, and so on.

The Poisson brackets $\{,\}_k$ defined by $\mathcal{J}_k$ for $k = 1, 2$ are compatible with each other, and (5.7) can be written in bi-Hamiltonian form:

$$u_t = \mathcal{J}_1 \frac{\delta H_1}{\delta u} = \mathcal{J}_2 \frac{\delta H_2}{\delta u}.$$

Furthermore, KdV has an infinite sequence of commuting symmetries given in terms of the *recursion operator* $\mathfrak{R} = \mathcal{J}_2 \mathcal{J}_1^{-1}$ by

$$u_{t_j} = \mathfrak{R}\, u_x, \quad j = 0, 1, 2, \ldots,$$

with a corresponding sequence of conserved quantities in involution with respect to both brackets.

If a conserved quantity is a local functional, given in the form $H[u] = \int \mathcal{H}\, dx$ for a density $\mathcal{H} = \mathcal{H}(u, u_x, u_{xx}, \ldots)$ given in terms of u and its derivatives, then its Fréchet derivative is given in terms of the Euler operator $\mathfrak{E}$:

$$\frac{\delta H}{\delta u} = \mathfrak{E} \cdot \mathcal{H} := \sum_{j=0}^{\infty} (-D_x)^j \frac{\partial \mathcal{H}}{\partial u_{jx}}.$$

Example 5.6 (Camassa–Holm equation). The equation

$$u_t - u_{xxt} = uu_{xxx} + 2u_x u_{xx} - 3uu_x \qquad (5.10)$$

has three simple conserved quantities given in terms of u:

$$H_0 = \int u\, dx, \quad H_1 = \int \frac{1}{2}(u_x^2 + u^2)\, dx, \quad H_2 = \int \frac{1}{2}(uu_x^2 + u^3)\, dx.$$

To write it in bi-Hamiltonian form [4], introduce the compatible pair

$$\mathcal{J}_1 = -(mD_x + D_x m), \quad \mathcal{J}_2 = D_x^3 - D_x, \quad \text{with } m = u - u_{xx}.$$

Then, since $u(x,t) = (1 - D_x^2)^{-1} m(x,t) = \frac{1}{2} \int_{\mathbb{R}} e^{-|x-y|} m(y,t)\, dy$, (5.10) is

$$m_t = \mathcal{J}_1 \frac{\delta H_1}{\delta m} = \mathcal{J}_2 \frac{\delta H_2}{\delta m}.$$

Remark 5.7 (Soliton solutions). Integrable Hamiltonian PDEs typically have exact solutions in the form of localized nonlinear waves called *solitons*, that preserve their speed and amplitude after collisions. The construction of solitons can be achieved using the inverse scattering problem for an associated linear system known as a *Lax pair* (see [2, 13]). For the KdV equation (5.7) the 1-soliton solution is

$$u(x,t) = 2k^2 \text{sech}^2(k(x - x_0) + 4k^3 t),$$

where $k > 0$ and x_0 are arbitrary parameters, and an N-soliton solution is a nonlinear superposition of N of these. The Camassa–Holm equation (5.10) has peaked solitons called *peakons*, which are weak solutions with

$$u(x,t) = \sum_{j=1}^{N} p_j(t) e^{-|x - q^j(t)|}, \quad m(x,t) = 2 \sum_{j=1}^{N} p_j(t) \delta(x - q^j(t)),$$

where the peak positions q^j and amplitudes p_j evolve according to a geodesic flow with Hamiltonian (4.7) and co-metric

$$g^{jk}(\mathbf{q}) = e^{-|q^j - q^k|}.$$

6. Geometry and Soliton Dynamics

In this section, we discuss the application of differential geometry to the dynamics of topological solitons. We also introduce some concepts in topology.

6.1. *Homotopy theory*

Given a manifold M and an interval $I = [0,1]$ we can define *paths*

$$\alpha : I \to M : t \mapsto \alpha(t), \quad \text{where } \alpha(0) = p_0, \ \alpha(1) = p_1.$$

A *loop* is a path with $p_0 = p_1$. Paths can be multiplied via

$$\alpha * \beta(s) = \begin{cases} \alpha(2s) & 0 \leq s \leq \frac{1}{2}, \\ \beta(2s - 1) & \frac{1}{2} \leq s \leq 1. \end{cases}$$

The constant path is $c(s) = p_0$ for all $s \in I$. The inverse of a paths is $\alpha^{-1}(s) = \alpha(1 - s)$. *This is not a group, yet!*

Definition 6.1 (Homotopy). Let $\alpha, \beta : I \to M$ be loops at p_0. The loops α and β are *homotopic*, denoted by $\alpha \sim \beta$, if there exists a continuous map $F : I \times I \to M$ such that $F(s,0) = \alpha(s)$ and $F(s,1) = \beta(s)$ for all $s \in I$. Furthermore, $F(0,t) = F(1,t) = p_0$ for all $t \in I$.

It can be shown that $\alpha \sim \beta$ is an equivalence relation. Let $[\alpha]$ be the equivalence class which contains α. Define a product on equivalence classes by $[\alpha] * [\beta] = [\alpha * \beta]$. This gives the *fundamental group* $\pi_1(M, p_0)$.[g]

Example 6.2. The winding number $\pi_1(S^1) = \mathbb{Z}$ counts how many times a path goes around the circle. Similarly, for the punctured plane $\pi_1(\mathbb{R}^2 \setminus \{0\}) = \mathbb{Z}$ counts how many times a path encircles the origin. The fundamental group of the torus is $\pi_1(T^2) = \mathbb{Z} \oplus \mathbb{Z}$; more generally, $\pi_1(M \times N) = \pi_1(M) \oplus \pi_1(N)$.

This generalizes naturally to higher homotopy groups: Consider maps from the cube $I^n = I \times \cdots \times I$ to a manifold M such that all the points on the boundary ∂I^n of the cube are mapped to $p_0 \in M$:

$$\alpha : (I^n, \partial I^n) \to (M, p_0).$$

Again we can form the product $\alpha * \beta$ and define the equivalence classes $[\alpha]$ (also known as homotopy classes), giving the nth homotopy group $\pi_n(M)$.

Remark 6.3 (Summary of important results).

- Homotopy groups are Abelian for $n > 1$, i.e. $[\alpha] * [\beta] = [\beta] * [\alpha]$.
- Manifolds M with $\pi_1(M) = 1$ are called *simply-connected.*
- The *degree* of a map is $\pi_n(S^n) = \mathbb{Z}$ related to the number of pre-images.
- $\pi_n(S^d) = 1$ for $1 \leq n < d$: the map is not onto, therefore contractible.
- $\pi_{n+1}(S^n) = \mathbb{Z}_2$, for $n \geq 3$, but $\pi_3(S^2) = \mathbb{Z}$ which is related to the Hopf bundle in Example 2.3.
- Homotopy groups of spheres are really complicated, e.g. $\pi_{n+2}(S^2) = \mathbb{Z}_2$ for $n \geq 2$.
- *Spectral sequences* are an important tool: Let G be a Lie group with subgroup H; then

$$\cdots \to \pi_n(H) \to \pi_n(G) \to \pi_n(G/H) \to \pi_{n-1}(H) \to \pi_{n-1}(G) \to \pi_{n-1}(G/H) \to \cdots$$

is a long exact sequence. Example: $G = S^3$, $H = S^1$ and $G/H = S^2$.

6.2. *Homotopy groups and field theory*

Field configurations at a fixed time are maps $\phi : \mathbb{R}^d \to M$, from flat space to a target space. Homotopies of maps occur naturally (e.g. time evolution is continuous and connects different field configurations in the same homotopy

---

[g]If M is arcwise connected then $\pi_1(M, p_0)$ is isomorphic to $\pi_1(M, p_1)$.

class). Two scenarios naturally give rise to homotopy groups, both arising from boundary conditions (due to finite energy):

(1) *One-point compactification:* There is a unique vacuum $v_0 \in M$, namely, $\phi(\mathbf{x}) = v_0$ for $\mathbf{x} \to \infty$. We can identify all these points, so that topologically $\mathbb{R}^d \cup \{\infty\} = S^d$. So we need $\pi_d(M)$.

(2) *Non-trivial maps at infinity:* The vacuum is degenerate and forms a submanifold N of M. Then in the limit $|\mathbf{x}| \to \infty$ there is a continuous map $\phi|_\infty : S_\infty^{d-1} \to N$. So we need $\pi_{d-1}(N)$.

This leads to the following classification of solitons.

$\pi_n(S^k)$	ungauged	gauged
$\pi_1(S^1)$	Kinks	**Vortices**
$\pi_2(S^2)$	Baby-Skyrmions, Lumps	Monopoles
$\pi_3(S^3)$	Skyrmions	Instantons
$\pi_3(S^2)$	Hopf Solitons	

6.3. *Ginzburg–Landau vortices*

The Ginzburg–Landau energy is given by (3.5). In coordinates $\mathbf{x} = (x^1, x^2) = (x, y)$ it is invariant under gauge transformations

$$\phi(\mathbf{x}) \mapsto e^{i\alpha(\mathbf{x})} \phi(\mathbf{x}), \quad a_i(\mathbf{x}) \mapsto a_i(\mathbf{x}) + \partial_i \alpha(\mathbf{x}),$$

where $e^{i\alpha(\mathbf{x})}$ is a spatially varying phase. The quantity

$$B \equiv f_{12} = \partial_1 a_2 - \partial_2 a_1$$

is the magnetic field. The vacuum is $\phi = 1$, $a_i = 0$ and gauge transformations of this, and we require $|\phi| \to 1$ as $|\mathbf{x}| \to \infty$.

Transforming to polar coordinates $(x, y) = (\rho \cos \theta, \rho \sin \theta)$, the energy is

$$V = \frac{1}{2} \int_0^\infty \int_0^{2\pi} \left(B^2 + \overline{D_\rho \phi} D_\rho \phi + \frac{1}{\rho^2} \overline{D_\theta \phi} D_\theta \phi + \frac{\lambda}{4} \left(1 - \bar{\phi} \phi \right)^2 \right) \rho \, d\rho \, d\theta.$$

For finite energy fields we can fix the gauge asymptotically, so that

$$\lim_{\rho \to \infty} \phi(\rho, \theta) = e^{i\alpha(\theta)},$$

where α is a continuous function of θ. As θ increases from 0 to 2π, $\alpha(\theta)$ increases by $2\pi N$ (ϕ is single valued). The *winding number* N is an arbitrary integer that cannot change under smooth deformations of the field, so remains constant in time, and is also invariant under smooth gauge transformations.

For finite energy, the covariant derivative $D_\theta \phi = (\partial_\theta - ia_\theta)\phi$ has to vanish as $\rho \to \infty$, so $\phi \sim e^{i\alpha(\theta)}$ implies $a_\theta = \frac{d\alpha}{d\theta}$. Hence, by Stokes' theorem,

$$\int_{\mathbb{R}^2} B \; d^2 x = \int_0^{2\pi} a_\theta \; d\theta|_{\rho \to \infty} = \alpha(2\pi) - \alpha(0) = 2\pi N, \qquad (6.1)$$

so N measures the units of magnetic flux in the plane. If ϕ has only isolated zeros, then the number of these (counted with multiplicity) is N — see Fig. 3(a). A zero of ϕ is said to have multiplicity k, if on a small circle enclosing it, $-\arg \phi$ increases by $2\pi k$. For simple zeros $k = \pm 1$.

Let E_N be the minimal energy V of N vortices. There are three different regimes: **(i) type I:** $\lambda < 1$, $E_N < NE_1$ — the vortices attract; **(ii) type II:** $\lambda > 1$, $E_N > NE_1$ — the vortices repel; **(iii) critical coupling:** $\lambda = 1$, $E_N = NE_1$ — no forces between static vortices. (See Fig. 3(b).) At critical coupling, by "completing the square" V can be written as

$$V = \frac{1}{2} \int \left(\left(B - \frac{1}{2}\left(1 - \bar{\phi}\phi\right) \right)^2 + \overline{(D_1\phi + iD_2\phi)}\,(D_1\phi + iD_2\phi) + B \right) d^2 x$$

$$\geq \pi N,$$

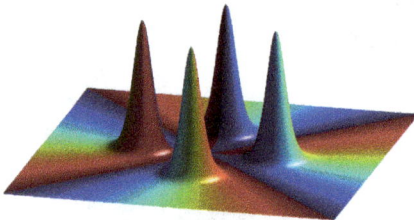
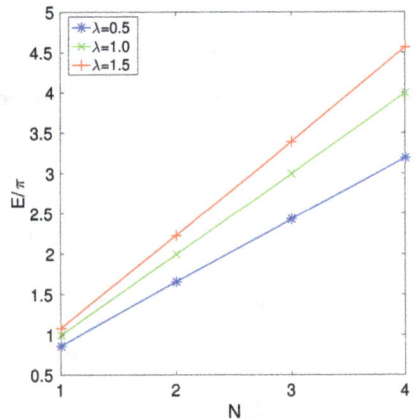

(a)

(b)

Fig. 3.   Figure (a) shows the energy density of a vortex configuration of charge $N = 4$. The four peaks correspond to four single vortices. The colour indicates the phase of the Higgs field ϕ. Clearly, the colour circle is completed four times as one goes around the "circle at infinity" once. Figure (b) shows the energy of Ginzburg–Landau vortices for charges $N = 1, \ldots, 4$ for coupling constant $\lambda = 0.5, 1$ and 1.5. For $\lambda = 1$ we observe $E_N = NE_1$.

using (6.1). Demanding that the squares vanish yields the Bogomolny equations

$$D_1\phi + iD_2\phi = 0, \quad B - \tfrac{1}{2}\left(1 - \bar{\phi}\phi\right) = 0. \tag{6.2}$$

These equations cannot be solved analytically. However, a lot is known about the solutions. For given topological charge N, the Bogomolny equations have a $2N$-dimensional manifold of static solutions, known as the *moduli space M_N*. (Gauge equivalent solutions are identified.) All zeros of ϕ have positive multiplicity (generically there are only simple zeros). A solution is completely determined by the locations of these zeros, which can be anywhere, so M_N is parametrized by N unordered points in $\mathbb{R}^2$, requiring $2N$ coordinates. There are no *static* forces between vortices for $\lambda = 1$, however, there will be *velocity-dependent* forces.

6.4. *Relativistic vortex dynamics*

In $2+1$ dimensions with $(x^\mu) = (t, \mathbf{x})$, the standard relativistic Lagrangian is

$$\mathcal{L} = \frac{1}{2}\overline{D_\mu\phi}D^\mu\phi - \frac{1}{4}f_{\mu\nu}f^{\mu\nu} - \frac{\lambda}{8}\left(1 - \bar{\phi}\phi\right)^2.$$

In the following, we will often use complex coordinates $z = x + iy$.

Parametrizing the moduli space for $\lambda = 1$ by vortex positions $z = Z_i$, assumed to be time dependent, gives a reduced Lagrangian for geodesics on M_N,

$$L_{\text{red.}} = \tfrac{1}{2}\left(g_{rs}\dot{Z}_r\dot{Z}_s + g_{r\bar{s}}\dot{Z}_r\dot{\bar{Z}}_s + g_{\bar{r}\bar{s}}\dot{\bar{Z}}_r\dot{\bar{Z}}_s\right) - \pi N, \tag{6.3}$$

summed for $1 \le r, s \le N$. Setting $h = \log|\phi|^2$ in (6.2) implies

$$\nabla^2 h + 1 - e^h = 4\pi \sum_{r=1}^{N} \delta^2(z - Z_r).$$

The δ functions arise because h has logarithmic singularities at the zeros Z_r of ϕ. Expanding h around the point Z_r gives

$$h(z, \bar{z}) = 2\log|z - Z_r| + a_r + \tfrac{1}{2}\bar{b}_r(z - Z_r) + \tfrac{1}{2}b_r(\bar{z} - \bar{Z}_r) + \cdots.$$

After a long calculation, we find $g_{rs} = 0 = g_{\bar{r}\bar{s}}$ in (6.3), so this purely kinetic Lagrangian gives rise to the moduli space metric

$$g = \pi\left(\delta_{rs} + 2\frac{\partial b_s}{\partial Z_r}\right)dZ_r d\bar{Z}_s,$$

which can be shown to be Kähler. This structure provides a lot of information about the metric, although it is only know implicitly. The moduli space
approximation captures the dynamics of vortices, in particular right-angle
scattering.

We can consider physical spaces X with a different metric, e.g.

$$ds^2 = dt^2 - \Omega(dx^2 + dy^2),$$

where $\Omega = \Omega(x, y)$ is a conformal factor defining the Riemannian metric
on X. Again we can "complete the square" and obtain the Bogomolny
equations

$$D_1\phi + iD_2\phi = 0, \quad B - \frac{\Omega}{2}\left(1 - \bar{\phi}\phi\right) = 0. \tag{6.4}$$

For a surface X with metric defined by Ω the integral

$$c_1 = \frac{1}{2\pi}\int_X f = \frac{1}{2\pi}\int_X B \, d^2x$$

is an integer. This topological invariant is known as the first *Chern number.*

If X has a finite area A then we can integrate the second Bogomolny
equation over X to obtain

$$2\int_X B \, d^2x + \int_X |\phi|^2 \Omega \, d^2x = \int_X \Omega \, d^2x \Longrightarrow A$$

$$= 4\pi N + \int_X |\phi|^2 \Omega \, d^2x \geq 4\pi N,$$

which is called the Bradlow limit. In other words, each vortex needs an area
of at least 4π. At the Bradlow bound $A = 4\pi N$ both equations (6.4) can
trivially be solved by $\phi = 0$ and $B = \frac{\Omega}{2}$. For the torus T^2 the moduli space
metric has been calculated as an expansion around the Bradlow limit. For
the sphere S^2 moduli space metric is essentially the Fubini–Study metric
close to the Bradlow limit.

For the hyperbolic plane, the background metric on the Poincaré disc
$|z| < 1$ is $ds^2 = \Omega \, dz \, d\bar{z}$, with the conformal factor $\Omega = 8(1 - |z|^2)^{-2}$.
Setting $h = \log|\phi|^2$ we can again derive an equation for h, namely

$$\nabla^2 h + \Omega - \Omega e^h = 4\pi \sum_{r=1}^{N} \delta^2(z - Z_r).$$

and this can be transformed to Liouville's equation, which is integrable. In
this case, the moduli space is known explicitly, and

$$\phi = \frac{1 - |z|^2}{1 - |f|^2}\frac{df}{dz}, \quad \text{where } f(z) = \prod_{i=1}^{N+1}\left(\frac{z - c_i}{1 - \bar{c}_i z}\right),$$

with $|c_i| < 1$. The positions of the vortices are the zeros of $\frac{df}{dz}$, and the moduli space metric is

$$g = \pi \left(\Omega(Z_r)\delta_{rs} + 2\frac{\partial b_s}{\partial Z_r} \right) dZ_r d\bar{Z}_s$$

but now we can calculate b_s for special cases. The metric for n vortices on a regular polygon with m vortices fixed at the origin is given by

$$ds^2 = \frac{4\pi n^3 |\alpha|^{2n-2} d\alpha \ d\bar{\alpha}}{(1 - |\alpha|^{2n})^2} \left(1 + \frac{2n \left(1 + |\alpha|^{2n}\right)}{\sqrt{(m+1)^2 \left(1 - |\alpha|^{2n}\right)^2 + 4n^2 |\alpha|^{2n}}} \right)$$

for $n \neq m + 1$, and by

$$ds^2 = \frac{12\pi n^3 |\alpha|^{2n-2} d\alpha \ d\bar{\alpha}}{(1 - |\alpha|^{2n})^2}$$

for $n = m+1$. The non-trivial zeros are at $z = \alpha \, e^{2\pi i k/n}$ for $k = 0, \ldots, n-1$.

7. Summary

In these notes we have tried to give a flavour of the most basic differential geometric ideas that are important in mathematical physics, and provide a dictionary between mathematical and physical terminology. The interested reader is encouraged to delve into the bibliography to gain a deeper understanding.

Acknowledgements

A. Hone acknowledges the support of EPSRC Fellowship EP/M004333/1. S. Krusch was supported by the EPSRC First Grant EP/I034491/1. S. Krusch wants to thank J. Ashcroft for creating Figures 3(a) and 3(b).

References

[1] V. I. Arnold, *Mathematical Methods of Classical Mechanics.* Graduate Texts in Mathematics, Vol. 60. Springer-Verlag (1978).
[2] O. Babelon, D. Bernard and M. Talon, *Introduction to Classical Integrable Systems.* Cambridge Monographs on Mathematical Physics. Cambridge University Press (2007).

[3] M. Błaszak, *Multi-Hamiltonian Theory of Dynamical Systems*. Springer (1998).

[4] R. Camassa and D. D. Holm, An integrable shallow water equation with peaked solitons. *Phys. Rev. Lett.* **71**, 1661–1664 (1993).

[5] Y. Choquet-Bruhat, *Géometrie Differentielle et Systèmes Extérieurs*. Dunod (1968).

[6] W. A. Fedak and J. J. Prentis, The 1925 Born and Jordan paper "On quantum mechanics". *Amer. J. Phys.* **77**, 128–139 (2009).

[7] T. Frankel, *The Geometry of Physics*, 3rd edn. Cambridge University Press (2012).

[8] D. S. Freed and K. K. Uhlenbeck (eds.), *Geometry and Quantum Field Theory*. IAS/Park City Mathematics Series, Vol. 1. American Mathematical Society (1995).

[9] I. M. Gelfand and S. V. Fomin, *Calculus of Variations*. Dover (2000).

[10] V. Guillemin and S. Sternberg, *Symplectic Techniques in Physics*. Cambridge University Press (1984).

[11] V. Guillemin and S. Sternberg, *Variations on a Theme by Kepler*. AMS Colloquium Publications, Vol. 42. American Mathematical Society (1990).

[12] K. C. Hannabuss, *An Introduction to Quantum Theory*. Oxford University Press (1997).

[13] N. J. Hitchin, G. B. Segal and R. S. Ward, *Integrable Systems: Twistors, Loop Groups, and Riemann Surfaces*. Oxford University Press (1999).

[14] D. D. Holm, *Geometric mechanics. Part II: Rotating, Translating and Rolling*, 2nd edn. Imperial College Press (2011).

[15] L. P. Hughston and K. P. Tod, *An Introduction to General Relativity*. London Mathematical Society Student Texts, Vol. 5. Cambridge University Press (1991).

[16] F. Mandl, *Statistical Physics*. Wiley (1971).

[17] N. J. Manton and P. M. Sutcliffe, *Topological Solitons*. Cambridge Monographs on Mathematical Physics. Cambridge University Press (2004).

[18] M. Nakahara, *Geometry, Topology and Physics*. Graduate Student Series in Physics, 2nd edn. Institute of Physics (2003).

[19] Y. Nambu, Generalized Hamiltonian dynamics. *Phys. Rev. D* **7**, 2405–2412 (1973).

[20] P. J. Olver, *Applications of Lie Groups to Differential Equations*. Graduate Texts in Mathematics, Vol. 107. Springer (1986).

[21] A. Sudbery, *Quantum Mechanics and the Particles of Nature: An Outline for Mathematicians*. Cambridge University Press (1986).

[22] P. Vanhaecke, *Integrable Systems in the Realm of Algebraic Geometry*. Lecture Notes in Mathematics, Vol. 1638, 2nd edn. Springer (2001).

[23] A. Weinstein, The local structure of Poisson manifolds. *J. Differential Geom.* **18**, 523–557 (1983).

[24] R. P. Woodard, The theorem of Ostrogradsky, preprint (2015), `arXiv:1506.02210v2`.

Chapter 2

Distributions, Fourier Transforms and Microlocal Analysis

Yuri Safarov*

*Department of Mathematics, King's College
London WC2R 2LS, UK*

The aim of these notes is to introduce students to the modern field of mathematics called microlocal analysis. The first two sections are a brief introduction to the theory of the Fourier transform and tempered distributions. The third section is devoted to pseudodifferential operators (PDOs). It contains definitions and proofs of some classical results on PDOs. Finally, in the last two sections we discuss some applications of PDOs in analysis and the theory of partial differential equations. Solutions to the exercises (which are embedded in the text) and references for further reading are given at the end of the notes.

Notation

- $\mathbb{R}$ and $\mathbb{C}$ are the sets of real and complex numbers, respectively.
- $\mathbb{R}^n$ denotes the n-dimensional Euclidean space.
- $\operatorname{supp} f$ denotes the support of the function f, that is, $\operatorname{supp} f$ is the closure of the set $\{x : f(x) \neq 0\}$.
- A multi-index α is a set of n non-negative integers, $\alpha := \{\alpha_1, \alpha_2, \dots, \alpha_n\}$.
- If α, β are multi-indices then $|\alpha| := \alpha_1 + \alpha_2 + \cdots + \alpha_n$, $\alpha! := \alpha_1! \, \alpha_2! \dots \alpha_n!$ and $\alpha + \beta := \{\alpha_1 + \beta_1, \alpha_2 + \beta_2, \dots, \alpha_n + \beta_n\}$.
- $x = (x_1, x_2, \dots, x_n)$, $y = (y_1, y_2, \dots, y_n)$, $\xi = (\xi_1, \xi_2, \dots, \xi_n)$ are elements of $\mathbb{R}^n$.

*Yuri Safarov prepared these notes in August 2013 when this volume was first proposed. He died on 2nd June 2015. He is very much missed by his colleagues and students, and we dedicate this volume to his memory. *The Editors.*

- If $x \in \mathbb{R}^n$ and α is a multi-index then $x^\alpha := x_1^{\alpha_1} x_2^{\alpha_2} \ldots x_n^{\alpha_n}$, $\partial_{x_k} := \frac{\partial}{\partial x_k}$, $\partial_x^\alpha := \partial_{x_1}^{\alpha_1} \partial_{x_2}^{\alpha_2} \ldots \partial_{x_n}^{\alpha_n}$, $D_{x_k} := -i\partial_{x_k}$ and $D_x^\alpha := (-i)^{|\alpha|} \partial_x^\alpha$ where $i = \sqrt{-1}$.
- $C^\infty(\mathbb{R}^n)$ is the linear space of all infinitely differentiable functions on $\mathbb{R}^n$.
- $C_0^\infty(\mathbb{R}^n)$ is the linear space of all infinitely differentiable functions on $\mathbb{R}^n$ with compact supports.

1. Fourier transform

1.1. *Schwartz space $\mathcal{S}(\mathbb{R}^n)$*

Definition 1.1. We say that $f \in \mathcal{S}(\mathbb{R}^n)$ if f is infinitely differentiable and

$$\|f\|_{\alpha,\beta} := \sup_{x \in \mathbb{R}^n} |x^\beta \partial_x^\alpha f(x)| < \infty \tag{1.1}$$

for all multi-indices α, β.

Obviously, $\mathcal{S}(\mathbb{R}^n)$ is a linear space which contains $C_0^\infty(\mathbb{R}^n)$. If $f \in \mathcal{S}(\mathbb{R}^n)$ then, for all multi-indices α and all positive integers k, we have

$$|\partial_x^\alpha f(x)| \le c_{\alpha,k} (1 + |x|)^{-k}$$

with some constants $c_{\alpha,k}$. In other words, the functions $f \in \mathcal{S}(\mathbb{R}^n)$ and all their derivatives decay faster than any negative power of $|x|$ as $|x| \to \infty$. Therefore these functions are said to be *rapidly decreasing*.

Example 1.2. The function $f(x) = e^{-|x|^2}$ belongs to $\mathcal{S}(\mathbb{R}^n)$.

Lemma 1.3. *If $f \in \mathcal{S}(\mathbb{R}^n)$ then $x^\beta \partial_x^\alpha f(x) \in \mathcal{S}(\mathbb{R}^n)$ and*

$$\|x^\beta \partial_x^\alpha f\|_{\alpha',\beta'} \le \text{const} \sum_{|\alpha''| \le |\alpha+\alpha'|, |\beta''| \le |\beta+\beta'|} \|f\|_{\alpha'',\beta''}$$

for all multi-indices $\alpha, \beta, \alpha', \beta'$.

Proof. It is obvious.   □

We shall need the following version of Taylor's formula.

Lemma 1.4. *Let m be a non-negative integer, $f \in \mathcal{S}(\mathbb{R}^n)$ and $y \in \mathbb{R}^n$ be a fixed point. If f and all its derivatives up to the order m vanish at y then there exist functions $h_\beta \in \mathcal{S}(\mathbb{R}^n)$ such that*

$$f(x) = \sum_{\beta\,:\,|\beta|=m+1} (x-y)^\beta h_\beta(x), \quad \forall x \in \mathbb{R}^n. \tag{1.2}$$

Proof. Let $\zeta \in C_0^\infty(\mathbb{R}^n)$ with $\zeta \equiv 1$ in a neighbourhood of the point y. Denote $f_1 = (1 - \zeta)f$ and $f_2 = \zeta f$. Obviously, the function $h(x) := |x - y|^{-2m-2} f_1(x)$ belongs to $\mathcal{S}(\mathbb{R}^n)$. We have

$$f_1(x) = |x - y|^{2m+2} h(x) = \sum_{\beta\,:\,|\beta|=m+1} (x - y)^\beta P_\beta(x - y)\, h(x),$$

where P_β are some polynomials. Thus, the function f_1 can be represented in the form (1.2). By Taylor's formula,

$$f_2(x) = \sum_{\beta\,:\,|\beta|=m+1} (x - y)^\beta\, \tilde{h}_\beta(x),$$

where $\tilde{h}_\beta$ are some infinitely smooth functions. If $\tilde{\zeta} \in C_0^\infty(\mathbb{R}^n)$ with $\tilde{\zeta} \equiv 1$ on $\operatorname{supp} \zeta$ then, multiplying both parts of the above identity by $\tilde{\zeta}$, we obtain the expansion (1.2) for f_2. $\qquad\square$

1.2. Convergence in the space $\mathcal{S}(\mathbb{R}^n)$

Definition 1.5. We say that a sequence $\{f_k\} \subset \mathcal{S}(\mathbb{R}^n)$ converges to $f \in \mathcal{S}(\mathbb{R}^n)$ in the space $\mathcal{S}(\mathbb{R}^n)$ and write $f_k \xrightarrow{\mathcal{S}} f$ if $\|f - f_k\|_{\alpha,\beta} \to 0$ as $k \to \infty$ for all multi-indices α, β.

The space $\mathcal{S}(\mathbb{R}^n)$ can be provided with a metric ρ such that $f_k \xrightarrow{\mathcal{S}} f$ if and only if $\rho(f, f_k) \to 0$. In particular, one can take

$$\rho(f, g) := \sum_{\alpha,\beta} \frac{\|f - g\|_{\alpha,\beta}}{(\alpha + \beta)!\,(1 + \|f - g\|_{\alpha,\beta})}.$$

1.3. Fourier transform in $\mathcal{S}(\mathbb{R}^n)$

Definition 1.6. Let $f \in \mathcal{S}(\mathbb{R}^n)$. The function

$$\hat{f}(\xi) := \mathcal{F}_{x\to\xi} f(x) := (2\pi)^{-n/2} \int_{\mathbb{R}^n} e^{-ix\cdot\xi} f(x)\, dx, \quad \xi \in \mathbb{R}^n, \tag{1.3}$$

is called the *Fourier transform* of f.

The Fourier transform is well defined whenever the integral on the right-hand side of (1.3) exists and is finite for all $\xi \in \mathbb{R}^n$. Obviously, this is true if $f \in \mathcal{S}(\mathbb{R}^n)$.

Remark 1.7. By means of (1.3) one can define the Fourier transform for functions f from the Lebesgue space $L_1(\mathbb{R}^n)$ (which contains $\mathcal{S}(\mathbb{R}^n)$ as a subspace). However, as we shall see later, the Fourier transform can be extended to $L_1(\mathbb{R}^n)$ and even more general classes of functions in a different, more elegant way.

Lemma 1.8. *For all $f \in \mathcal{S}(\mathbb{R}^n)$ and all multi-indices α, we have*

$$\mathcal{F}_{x \to \xi}(D_x^\alpha f(x)) = \xi^\alpha \hat{f}(\xi), \quad \mathcal{F}_{x \to \xi}(x^\alpha f(x)) = (-1)^{|\alpha|} D_\xi^\alpha \hat{f}(\xi). \quad (1.4)$$

Proof.   The identities (1.4) are proved by integration by parts and differentiation under the integral sign. $\square$

Corollary 1.9. *If $f \in \mathcal{S}(\mathbb{R}^n)$, then $\hat{f} \in \mathcal{S}(\mathbb{R}^n)$, and the map $\mathcal{F} : f \to \hat{f}$ is continuous in the space $\mathcal{S}(\mathbb{R}^n)$.*

Proof.   For all $g \in \mathcal{S}(\mathbb{R}^n)$ we have

$$\sup_{\xi \in \mathbb{R}^n} |\hat{g}(\xi)| = (2\pi)^{-n/2} \sup_{\xi \in \mathbb{R}^n} \left| \int e^{ix \cdot \xi} g(x) \, dx \right| \leq (2\pi)^{-n/2} \int |g(x)| \, dx. \tag{1.5}$$

Therefore the corollary follows from Lemmas 1.3 and 1.8. $\square$

Example 1.10. Let us calculate the Fourier transform of the function $f(x) = \exp\left(-|x|^2/2\right)$. First, we consider the function $f_0(t) = \exp\left(-t^2/2\right)$ on $\mathbb{R}$. This function is a solution of the differential equation

$$f_0'(t) = -t \, f_0(t). \tag{1.6}$$

Applying the (one-dimensional) Fourier transform to (1.6) and taking into account (1.4), we obtain

$$it \, \hat{f}_0(t) = -i \, \hat{f}_0'(t),$$

where $\hat{f}_0'$ is the derivative of the Fourier transform $\hat{f}_0$. Now we see that

$$\left(\frac{\hat{f}_0(t)}{f_0(t)} \right)' = \frac{\hat{f}_0'(t) \, f_0(t) - \hat{f}_0(t) \, f_0'(t)}{f_0^2(t)} = \frac{-t \, \hat{f}_0(t) \, f_0(t) + t \, \hat{f}_0(t) \, f_0(t)}{f_0^2(t)} = 0,$$

which implies $\hat{f}_0(t) = c_0 \, f_0(t) = c_0 \exp\left(-t^2/2\right)$ with some constant $c_0 \geq 0$. Passing to the polar coordinates, we obtain

$$c_0^2 = (\hat{f}_0(0))^2 = (2\pi)^{-1} \left(\int e^{-t^2/2} \, dt \right)^2 = (2\pi)^{-1} \iint e^{-(t^2+\tau^2)/2} \, dt \, d\tau$$

$$= (2\pi)^{-1} \int_0^\infty \int_{\mathbf{S}^1} e^{-r^2/2} \, r \, d\theta \, dr = \int_0^\infty e^{-r^2/2} \, r \, dr = \frac{1}{2} \int_0^\infty e^{-s/2} \, ds = 1,$$

so $\hat{f}_0(t) = f_0(t) = \exp(-t^2/2)$. Finally,

$$(2\pi)^{-n/2} \int e^{-ix\cdot\xi} e^{-|x|^2/2} \, dx = \hat{f}_0(\xi_1)\,\hat{f}_0(\xi_2)\ldots\hat{f}_0(\xi_n) = \exp(-|\xi|^2/2).$$

Thus, $\hat{f}(\xi) = f(\xi) = \exp(-|\xi|^2/2)$.

1.4. *Inversion formula*

Theorem 1.11. *Let* $T : S(\mathbb{R}^n) \to S(\mathbb{R}^n)$ *be a linear map commuting with multiplication by* x_k *and differentiation* D_{x_k} *for all* $k = 1, \ldots, n$, *that is,*

$$T(x_k f) = x_k\,(Tf), \quad T(D_{x_k} f) = D_{x_k}(Tf), \quad \forall\, k = 1, 2, \ldots, n, \qquad (1.7)$$

for all $f \in S(\mathbb{R}^n)$. *Then there exists a constant* c *such that* $Tf = c\,f$ *for all* $f \in S(\mathbb{R}^n)$.

Proof. Let $f, g \in S(\mathbb{R}^n)$ and $y \in \mathbb{R}^n$ be a fixed point. If $f(y) = g(y)$ then, by Lemma 1.4,

$$f(x) - g(x) = \sum_{k=1}^{n}(x_k - y_k)\,h_k(x),$$

where $h_k \in S(\mathbb{R}^n)$. Now the first identity (1.7) implies that $(Tf)(y) = (Tg)(y)$. Thus, the value of Tf at any point y depends only on the value of f at the point y. Since T is a linear map, this implies that $(Tf)(y) = c(y)\,f(y)$, where $c(y)$ is some constant depending on y.

Since Tf is an infinitely differentiable function for every $f \in S(\mathbb{R}^n)$, the constant $c(y)$ smoothly depends on y. Applying the second identity (1.7), we obtain

$$c(y)\,\partial_{y_k} f(y) = \partial_{y_k}\,(c(y)\,f(y)) = f(y)\,\partial_{y_k} c(y) + c(y)\,\partial_{y_k} f(y)$$

for all $f \in S(\mathbb{R}^n)$ and $k = 1, 2, \ldots, n$. Therefore all first derivatives of c are identically equal to zero, which means that c does not depend on y. $\qquad\square$

Let $Jf(x) := f(-x)$. Obviously, J is a continuous operator in $S(\mathbb{R}^n)$ and $J\mathcal{F} = \mathcal{F}J$ (the latter is proved by changing variables $\xi = -\eta$ in (1.3)).

Corollary 1.12. *If* $f \in S(\mathbb{R}^n)$ *then*

$$f(x) = J(\mathcal{F}_{\xi\to x}\hat{f}(\xi)) = (2\pi)^{-n/2} \int e^{ix\cdot\xi}\,\hat{f}(\xi)\,d\xi. \qquad (1.8)$$

Proof.   Define $(Tf)(x) := J(\mathcal{F}_{\xi \to x}\hat{f}(\xi))$. In view of Corollary 1.9, T is a linear operator from $\mathcal{S}(\mathbb{R}^n)$ into $\mathcal{S}(\mathbb{R}^n)$. Lemma 1.8 implies that

$$T(D_{x_k}f) = J(\mathcal{F}_{\xi \to x}(\xi_k \hat{f}(\xi))) = J(-D_{x_k}(\mathcal{F}_{\xi \to x}\hat{f}(\xi))) = D_{x_k}(Tf),$$

$$T(x_k f) = J(\mathcal{F}_{\xi \to x}(-D_{x_k}\hat{f}(\xi))) = J(-x_k(\mathcal{F}_{\xi \to x}\hat{f}(\xi))) = x_k(Tf).$$

Therefore, by Theorem 1.11, there exists a constant c such that $Tf = cf$ for all $f \in \mathcal{S}(\mathbb{R}^n)$. If $f(x) = \exp(-|x|^2/2)$ then

$$c\,f(x) = (Tf)(x) = J(\mathcal{F}_{\xi \to x}\hat{f}(\xi)) = f(x)$$

(see Example 1.10). This implies that $c = 1$.   □

The linear operator

$$f(x) \to J(\mathcal{F}_{x \to \xi}f(x)) = (2\pi)^{-n/2}\int e^{ix\cdot\xi}\,\hat{f}(x)\,dx$$

is called the *inverse Fourier transform*. By Corollary 1.9, $J\mathcal{F}f \in \mathcal{S}(\mathbb{R}^n)$ and, by (1.8), $J\mathcal{F}\mathcal{F}f = \mathcal{F}J\mathcal{F}f = f$ whenever $f \in \mathcal{S}(\mathbb{R}^n)$. Thus, we have proved the following theorem.

Theorem 1.13. *The Fourier transform $\mathcal{F}$ is a one-to-one map from $\mathcal{S}(\mathbb{R}^n)$ onto $\mathcal{S}(\mathbb{R}^n)$ and $\mathcal{F}^{-1} = J\mathcal{F} = \mathcal{F}J$.*

Corollary 1.14 (Parseval's formula). *If $f, g \in \mathcal{S}(\mathbb{R}^n)$ then*

$$\int f(x)\,\overline{g(x)}\,dx = \int \hat{f}(x)\,\overline{\hat{g}(x)}\,d\pi x.$$

Proof.   Parseval's formula follows from (1.8) and the obvious identities

$$\overline{\hat{g}(x)} = \mathcal{F}^{-1}(\bar{g}), \quad \int \hat{f}(x)\,g(x)\,dx = \int f(x)\,\hat{g}(x)\,dx, \quad \forall f, g \in \mathcal{S}(\mathbb{R}^n). \tag{1.9}$$

Thus the proof is complete   □

2.   Tempered Distributions

2.1.   *Definition and examples*

A map $u : \mathcal{S}(\mathbb{R}^n) \to \mathbb{C}$ is said to be a functional on $\mathcal{S}(\mathbb{R}^n)$. The value of functional u on the function $f \in \mathcal{S}(\mathbb{R})$ is denoted by $\langle u, f \rangle$. We say that the functional u is linear if

$$\langle u, c_1 f + c_2 g \rangle = c_1 \langle u, f \rangle + c_2 \langle u, g \rangle, \quad \forall f, g \in \mathcal{S}(\mathbb{R}^n), \quad \forall c_1, c_2 \in \mathbb{C},$$

and u is continuous if $\langle u, f_k \rangle \to \langle u, f \rangle$ whenever $f_k \overset{S}{\to} f$.

Definition 2.1. A linear continuous functional on $\mathcal{S}(\mathbb{R}^n)$ is said to be a *tempered distributions*.

If u, v are tempered distribution and $c_1, c_2 \in \mathbb{C}$, let us define the distribution $c_1 u + c_2 v$ by

$$\langle c_1 u + c_2 v, f \rangle = c_1 \langle u, \varphi \rangle + c_2 \langle v, f \rangle, \quad \forall f \in \mathcal{S}(\mathbb{R}^n).$$

Then the set of tempered distributions becomes a linear space. This space is denoted by $\mathcal{S}'(\mathbb{R}^n)$.

Example 2.2. Let u be a polynomially bounded function on $\mathbb{R}^n$. If u is sufficiently nice (continuous, piecewise continuous or, more generally, measurable) then the functional defined by

$$\langle u, f \rangle := \int u(x) f(x) \, dx, \quad \forall f \in \mathcal{S}(\mathbb{R}^n), \tag{2.1}$$

is a tempered distribution. This allows us to identify the "regular" polynomially bounded functions with distributions. Obviously, two functions u_1 and u_2 define the same distribution then $u_1 = u_2$ "almost everywhere" (with respect to the Lebesgue measure). Further on we shall use the same notation u for the function on $\mathbb{R}^n$ and the corresponding distribution.

If u is not polynomially bounded then the integrals on the right-hand side of (2.1) may not converge in the usual sense. However, in many cases one can use a suitable regularization of these integrals in order to define a distribution generated by u. This distribution may, of course, depend on the choice of regularization.

Example 2.3. Let $x \in \mathbb{R}^n$ be a fixed point. The tempered distribution δ_x defined by

$$\langle \delta_x, f \rangle = f(x), \quad \forall f \in \mathcal{S}(\mathbb{R}^n),$$

is said to be the δ-function at x. The δ-function at the origin is usually denoted by δ or $\delta(y)$, where y indicates that δ is considered as a functional on the space of functions depending on the variables y.

Theorem 2.4. *A linear functional u on $\mathcal{S}(\mathbb{R}^n)$ is continuous if and only if there exists a constant C and a non-negative integer m such that*

$$|\langle u, f \rangle| \leq C \sum_{|\alpha + \beta| \leq m} \|f\|_{\alpha, \beta}, \quad \forall f \in \mathcal{S}(\mathbb{R}^n).$$

Proof. Assume first that there exist constants C and m for which the above estimate holds. If $f_j \xrightarrow{S} f$ then $\|f - f_j\|_{\alpha,\beta} \to 0$ for all α, β and, consequently, $\langle u, f - f_j \rangle \to 0$. This implies that u is continuous.

Conversely, let us assume that such constants C and m do not exist. Then there is a sequence of functions $f_j \in \mathcal{S}(\mathbb{R}^n)$ such that

$$|\langle u, f_j \rangle| > j \sum_{|\alpha+\beta| \leq j} \|f_j\|_{\alpha,\beta}.$$

If $g_j(x) = (\langle u, f_j \rangle)^{-1} f_j(x)$ then $\|g_j\|_{\alpha,\beta} \leq j^{-1}$ for all $j \geq |\alpha + \beta|$. This implies that $\|g_j\|_{\alpha,\beta} \to 0$ as $j \to \infty$ for all multi-indices α and β or, in other words, that $g_j \xrightarrow{S} 0$. On the other hand, $\langle u, g_j \rangle = 1$ for all j, so u is not a continuous functional on $\mathcal{S}(\mathbb{R}^n)$. □

Exercise 1. Prove that the functional $f \mapsto \lim_{\varepsilon \to 0} \int_{|t| > \varepsilon} t^{-1} f(t)\, dt$ belongs to $\mathcal{S}'(\mathbb{R})$.

Exercise 2. Find all distributions $u \in \mathcal{S}'(\mathbb{R})$ such that $\langle u, f \rangle = \int_{-\infty}^{\infty} t^{-1} f(t)\, dt$ for all functions $f \in \mathcal{S}(\mathbb{R})$ satisfying the condition $f(0) = 0$.

2.2. *Operators in the space of distributions*

Let A be a linear operator acting in the space $\mathcal{S}(\mathbb{R}^n)$.

Condition 2.5. There exists a continuous operator $A^T : \mathcal{S}(\mathbb{R}^n) \mapsto \mathcal{S}(\mathbb{R}^n)$ such that

$$\int (Au)(x)\, f(x)\, dx = \int u(x)\, (A^T f)(x)\, dx, \quad \forall u, f \in \mathcal{S}(\mathbb{R}^n). \tag{2.2}$$

Lemma 2.6. *If Condition 2.5 is fulfilled, then one can extend A to the space $\mathcal{S}'(\mathbb{R}^n)$.*

Proof. If $u \in \mathcal{S}'(\mathbb{R}^n)$, we define Au by

$$\langle Au, f \rangle := \langle u, A^T f \rangle, \quad \forall f \in \mathcal{S}(\mathbb{R}^n). \tag{2.3}$$

One can easily see that, under Condition 2.5, Au is a tempered distribution. If $u \in \mathcal{S}(\mathbb{R}^n)$ then (2.3) turns into (2.2), so (2.3) defines the same operator A on the space $\mathcal{S}(\mathbb{R}^n)$. □

Lemma 2.6 justifies the following definitions.

Definition 2.7. Let h be an infinitely smooth function on $\mathbb{R}^n$ which is polynomially bounded with all its derivatives. If $u \in \mathcal{S}'(\mathbb{R}^n)$ then hu is the distribution defined by

$$\langle hu, f \rangle := \langle u, hf \rangle, \quad \forall f \in \mathcal{S}(\mathbb{R}^n).$$

Definition 2.8. If $u \in \mathcal{S}'(\mathbb{R}^n)$ and α is a multi-index then $\partial_x^\alpha u$ is the distribution defined by

$$\langle \partial_x^\alpha u, f \rangle := (-1)^{|\alpha|} \langle u, \partial_x^\alpha f \rangle, \quad \forall f \in \mathcal{S}(\mathbb{R}^n).$$

In the same manner one can define other operators in $\mathcal{S}'(\mathbb{R}^n)$, in particular, the change of variables operator $u(x) \to v(x) = u(\tilde{x}(x))$ where $\tilde{x}(x)$ is a smooth vector function satisfying certain conditions at infinity.

Example 2.9. Let $x \in \mathbb{R}^n$ be a fixed point and $A_x f(y) := f(x - y)$. Then A_x is a continuous operator in $\mathcal{S}(\mathbb{R}^n)$ and $A_x^T = A_x$. If $u(y)$ is a distribution (y means that we apply u to functions depending on y) then, by (2.3),

$$\langle u(x - y), f(y) \rangle := \langle u(y), f(x - y) \rangle, \quad \forall f \in \mathcal{S}(\mathbb{R}^n).$$

In particular, for the δ-function (see Example 2.3) we have

$$\langle \delta(x - y), f(y) \rangle := \langle \delta(y), f(x - y) \rangle = f(x) = \langle \delta_x, f \rangle, \quad \forall f \in \mathcal{S}(\mathbb{R}^n),$$

that is, $\delta_x(y) = \delta(x - y)$. In a similar way one can show that $\delta_x(y) = \delta(y - x)$ which implies that $\delta(x - y) = \delta(y - x)$.

Example 2.10. Let $u(t)$ be the characteristic function of the positive half-line. Then, for every $s \in \mathbb{R}$, the derivative of the function $u(t - s)$ coincides with $\delta(t - s)$. For the second derivative of $u(t - s)$ we have

$$\langle u''(t - s), f(t) \rangle = \langle \delta'(t - s), f(t) \rangle = -f'(s), \quad \forall f \in \mathcal{S}(\mathbb{R}^n).$$

The distribution $\delta'(t - s)$ cannot be described in any simpler way. It is called the derivative of the δ-function at the point s.

Exercise 3. Let $-\infty < a_1 < a_2 < \cdots < a_m < \infty$ and u be a function on $\mathbb{R}$ with the following properties:

(1) u vanishes outside the interval $[a_1, a_m]$;
(2) u is continuously differentiable on every interval (a_k, a_{k+1});
(3) u has finite left and right limits at the points a_k.

Evaluate the derivative $u' \in \mathcal{S}'(\mathbb{R})$ of the function u.

2.3.   Supports of distributions

Generally speaking, a distribution does not take any particular value at one
fixed point. However, two distributions may coincide on an open set.

Definition 2.11. We say that the distribution u vanishes on an open set
Ω and write $u|_\Omega = 0$ if $\langle u, f \rangle = 0$ for all $f \in \mathcal{S}(\mathbb{R}^n)$ with supp $f \subset \Omega$. We
say that u coincides with another distribution v on Ω if $(u - v)|_\Omega = 0$.

In particular, the distribution u coincides with a function v on Ω if
$\langle u, f \rangle = \int_\Omega vf \, dx$ whenever supp $f \subset \Omega$.

Definition 2.12. If $u \in \mathcal{S}'(\mathbb{R}^n)$ then supp $u := \mathbb{R}^n \setminus \Omega_u$, where Ω_u is the
union of all open sets Ω such that $u|_\Omega = 0$.

Example 2.13. The support of any derivative of the δ-function at y coincides with the point y.

The support of a continuous function u coincides with the support of
the corresponding distribution (if u is not continuous then this statement is
correct modulo a set of Lebesgue measure zero). If h is a function satisfying
conditions of Definition 2.7 then

$$\text{supp} \, (hu) \subset (\text{supp} \, h) \cap (\text{supp} \, u), \quad \forall u \in \mathcal{S}'(\mathbb{R}^n).$$

In particular, if $h = 0$ in a neighbourhood of supp u then $hu = 0$. This is
not necessarily true if $h = 0$ only on supp u.

Example 2.14. If $h = 0$ at the origin then $h(x)\delta(x) \equiv 0$. However,

$$\langle h(x) \, \partial_{x_k} \delta(x), f(x) \rangle = - \, \partial_{x_k}(h(x)f(x))|_{x=0} = -h_{x_k}(0) \, f(0),$$

that is, $h(x) \, \partial_{x_k} \delta(x) = -(\partial_{x_k} h(0)) \, \delta(x)$.

The set of distributions with compact supports is denoted by $\mathcal{E}'(\mathbb{R}^n)$.
Theorem 2.4 implies the following result (see [2, Theorem 4.4.7]).

Theorem 2.15. If $u \in \mathcal{E}'(\mathbb{R}^n)$ then there exists a non-negative integer m
such that

$$u(x) = \sum_{|\alpha| \leq m} \partial_x^\alpha u_\alpha(x), \tag{2.4}$$

where u_α are some continuous compactly supported functions.

One can always choose C_0^∞-functions ψ_j such that $\sum_j \psi_j(x) \equiv 1$ (it is
called a partition of unity). Then an arbitrary distribution $u \in \mathcal{S}'(\mathbb{R}^n)$ is

represented as the sum of distributions $u_j = \psi_j u$ with compact supports, that is, as a sum of distributions of the form (2.4).

2.4. Fourier transform in $\mathcal{S}'(\mathbb{R}^n)$

By Corollary 1.9, the Fourier transform $\mathcal{F}$ and the inverse Fourier transform $\mathcal{F}^{-1} = J\mathcal{F}$ are linear continuous operators in $\mathcal{S}(\mathbb{R}^n)$. Obviously, $\mathcal{F}$ and $\mathcal{F}^{-1}$ satisfy Condition 2.5 with $\mathcal{F}^T = \mathcal{F}$ and $(\mathcal{F}^{-1})^T = \mathcal{F}^{-1}$. Therefore, according to Lemma 2.6, the operators $\mathcal{F}$ and $\mathcal{F}^{-1}$ can be extended to $\mathcal{S}'(\mathbb{R}^n)$.

Definition 2.16. If $u \in \mathcal{S}'(\mathbb{R}^n)$ then $\hat{u} = \mathcal{F}u$ and $\mathcal{F}^{-1}u$ are the tempered distributions defined by

$$\langle \mathcal{F}u, f \rangle := \langle u, \mathcal{F}f \rangle, \quad \langle \mathcal{F}^{-1}u, f \rangle := \langle u, \mathcal{F}^{-1}f \rangle, \quad \forall f \in \mathcal{S}(\mathbb{R}^n).$$

Lemma 1.8, Theorem 1.13 and Definition 2.16 immediately imply the following lemma.

Lemma 2.17. *For all* $u \in \mathcal{S}'(\mathbb{R}^n)$ *we have* $\mathcal{F}^{-1}\mathcal{F}u = \mathcal{F}\mathcal{F}^{-1}u = u$ *and*

$$\mathcal{F}_{x \to \xi}(D_x^\alpha u) = \xi^\alpha \, \hat{u}(\xi), \quad \mathcal{F}_{x \to \xi}(x^\alpha u) = (-1)^{|\alpha|} D_\xi^\alpha \hat{u}(\xi).$$

Example 2.18. Let u be a "nice" polynomially bounded function (as in Example 2.2). Then

$$\langle \hat{u}, f \rangle = \langle u, \hat{f} \rangle = (2\pi)^{-n/2} \int u(x) \left(\int e^{-ix \cdot \xi} f(\xi) \, d\xi \right) dx, \quad \forall f \in \mathcal{S}(\mathbb{R}^n).$$

If we can change the order of integration then

$$\langle \hat{u}, f \rangle = (2\pi)^{-n/2} \int \left(\int e^{-ix \cdot \xi} u(x) \, dx \right) f(\xi) \, d\xi, \quad \forall f \in \mathcal{S}(\mathbb{R}^n),$$

which implies that $\hat{u}(\xi) = (2\pi)^{-n/2} \int e^{-ix \cdot \xi} u(x) \, dx$. In particular, this formula holds for all functions u from the Lebesgue space $L_1(\mathbb{R}^n)$ (see Remark 1.7).

Example 2.19. If $\delta(x)$ is the δ-function then $\mathcal{F}_{x \to \xi}(D_x^\alpha \delta(x)) = (2\pi)^{-n/2} \xi^\alpha$. Indeed,

$$\langle \mathcal{F}_{x \to \xi}(D_x^\alpha \delta(x)), f(\xi) \rangle = (-1)^{|\alpha|} \langle \delta(x), D_x^\alpha \hat{f}(x) \rangle = (2\pi)^{-n/2} \int \xi^\alpha \, f(\xi) \, d\xi.$$

2.5.   *Divergent integrals*

We have defined the Fourier transform for all distributions $u \in \mathcal{S}'(\mathbb{R}^n)$, in particular, for all continuous polynomially bounded functions u. This means, in fact, that we have defined the integral $\int e^{-ix\cdot\xi} u(x)\,dx$ for every such a function u. Of course, this integral may not converge in the classical sense, but can be understood as a distribution in ξ. This idea can be generalized as follows.

Definition 2.20. Let $z \in \mathbb{R}^N$, $\xi \in \mathbb{R}^n$ and $G(z,\xi)$ be a continuous polynomially bounded function on $\mathbb{R}^N \times \mathbb{R}^n$. We shall say that the integral $\int G(z,\xi)\,d\xi$ converges in the sense of distributions if the consecutive integral $\int \left(\int G(z,\xi) f(z)\,dz \right) d\xi$ converges for every $f \in \mathcal{S}(\mathbb{R}^N)$ and the linear functional $\int G(z,\xi)\,d\xi$ defined by

$$\left\langle \int G(z,\xi)\,d\xi \,,\, f \right\rangle := \int \left(\int G(z,x)\,f(z)\,dz \right) dx, \quad \forall f \in \mathcal{S}(\mathbb{R}^N), \quad (2.5)$$

belongs to $\mathcal{S}'(\mathbb{R}^N)$.

Considering the integral $\int G(z,\xi)\,d\xi$ as a distribution, one can operate with it as with an absolutely convergent integral: formally integrate by parts, differentiate under the integral sign, etc. A rigorous justification of all these operations is obtained with the use of Definition 2.20.

Example 2.21. We have $(2\pi)^{-n} \int e^{-ix\cdot\xi}\,d\xi = \delta(x)$. Indeed, if $f \in \mathcal{S}(\mathbb{R}^n)$ then

$$(2\pi)^{-n} \int \left(\int e^{-ix\cdot\xi} f(x)\,dx \right) d\xi = (2\pi)^{-n/2} \int \hat{f}(\xi)\,d\xi$$

$$= \mathcal{F}^{-1}_{\xi \to y}\hat{f}(\xi)|_{y=0} = f(0).$$

It may well happen that the distribution $\int G(z,\xi)\,d\xi$ coincides with a function even if the integral does not converge in the usual sense.

Example 2.22. For all non-zero $z \in \mathbb{C}$ with $\operatorname{Re} z \leq 0$ we have

$$\mathcal{F}_{x\to\xi} \exp(z|x|^2/2) = (2\pi)^{-n/2} \int e^{z|x|^2/2}\, e^{-ix\cdot\xi}\,dx = z^{-n/2} \exp(z^{-1}|\xi|^2/2),$$
$$(2.6)$$

where $z^{-n/2} = |z|^{-n/2} \exp(-\frac{n}{2} \arg z)$ and $\arg z \in [-\pi/2, \pi/2]$. In particular, for $z = i$,

$$\mathcal{F}_{x\to\xi} \exp(i|x|^2/2) = (2\pi)^{-n/2} \int e^{i|x|^2/2 - ix\cdot\xi}\,dx = e^{i\pi n/4}e^{-i|\xi|^2/2}.$$

Proof. If $z = 1$ then $(2\pi)^{-n/2} \int e^{-|y|^2/2} e^{-iy\cdot\xi} \, dy = \exp(-|\xi|^2/2)$ (see Example 1.10). Changing variables $y = |z|^{1/2} x$, we see that (2.6) holds for all real negative z.

Let us fix an arbitrary complex number z_0 with $\operatorname{Re} z_0 < 0$, substitute $e^{-z|x|^2/2} = e^{-(z-z_0)|x|^2/2} e^{-z_0|x|^2/2}$ and expand the function $e^{-(z-z_0)|x|^2/2}$ into its Taylor series at the point $z = z_0$. Integrating term by term the series obtained, we see that for each fixed $\xi \in \mathbb{R}^n$ the function $\mathcal{F}_{x\to\xi} \exp(-z|x|^2/2)$ is given by an absolutely convergent power series in a neighbourhood of z_0. This implies that $\mathcal{F}_{x\to\xi} \exp(-z|x|^2/2)$ is analytic in the open half-plane $\{z \in \mathbb{C} : \operatorname{Re} z < 0\}$. The function $z^{-n/2} \exp(-z^{-1}|\xi|^2/2)$ is also analytic in this half-plane and, by the above, coincides with $\mathcal{F}_{x\to\xi} \exp(-z|x|^2/2)$ on the negative half-line. Now, from the identity theorem for analytic functions, it follows that (2.6) holds for all z with $\operatorname{Re} z < 0$.

Finally, letting $\operatorname{Re} z \to 0$, we obtain (2.6) for all imaginary numbers $z \neq 0$. $\qquad\square$

3. Schwartz Kernels, Oscillatory Integrals and Pseudodifferential Operators

3.1. *Schwartz kernels*

Theorem 3.1. *For every linear continuous operator A in the space $\mathcal{S}(\mathbb{R}^n)$ there exists a family of tempered distributions $\mathcal{A}(x, \cdot)$ depending on the parameter $x \in \mathbb{R}^n$ such that*

$$Av(x) = \langle \mathcal{A}(x, y), v(y) \rangle, \quad \forall x \in \mathbb{R}^n.$$

Proof. For every $x \in \mathbb{R}^n$ the map $v \to Av(x)$ is a linear continuous functional on $\mathcal{S}(\mathbb{R}^n)$, that is, a tempered distribution which we denote $\mathcal{A}(x, \cdot)$. $\qquad\square$

It is clear from the proof that $\mathcal{A}$ is uniquely defined by the operator A.

Definition 3.2. The family of distributions $\mathcal{A}$ is said to be the *Schwartz kernel* of the operator A.

If A is a linear continuous operator in $\mathcal{S}(\mathbb{R}^n)$ then, for every $u \in \mathcal{S}'(\mathbb{R}^n)$, the map

$$v(x) \to \langle u(x), Av(x) \rangle = \langle u(x), \langle \mathcal{A}(x, y), v(y) \rangle \rangle, \quad v \in \mathcal{S}(\mathbb{R}^n),$$

is a tempered distribution.

Definition 3.3. The linear operator A^T in $\mathcal{S}'(\mathbb{R}^n)$ defined by

$$\langle A^T u(x), v(x) \rangle = \langle u(x), Av(x) \rangle, \quad \forall u \in \mathcal{S}'(\mathbb{R}^n), \ \forall v \in \mathcal{S}(\mathbb{R}^n),$$

is said to be the *transposed operator* to A.

Now Condition 2.5 be rewritten as follows.

Condition 3.4. The transposed operator A^T continuously maps $\mathcal{S}(\mathbb{R}^n)$ into itself.

If Condition 3.4 is fulfilled then A^T also has a Schwartz kernel $\mathcal{A}^T(x, y)$.

Lemma 3.5. *Let A be a linear continuous operator in $\mathcal{S}(\mathbb{R}^n)$. If its Schwartz kernel $\mathcal{A}$ can be considered as a distribution $\mathcal{A}(\cdot, y)$ smoothly depending on $y \in \mathbb{R}^n$, that is, if there exists a family of distributions $\mathcal{A}(\cdot, y)$ such that $\langle \mathcal{A}(x, y), u(x) \rangle \in \mathcal{S}(\mathbb{R}^n)$ and*

$$\int \langle \mathcal{A}(x, y), u(x) \rangle v(y) \, dy = \int u(x) \langle \mathcal{A}(x, y), v(y) \rangle \, dx \qquad (3.1)$$

for all $u, v \in \mathcal{S}(\mathbb{R}^n)$, then A satisfies Condition 3.4 and $\mathcal{A}^T(x, y) = \mathcal{A}(y, x)$.

Proof. The identity (3.1) implies that

$$\langle Bu(x), v(x) \rangle = \langle u(x), Av(x) \rangle = \langle A^T u(x), v(x) \rangle, \quad \forall u, v \in \mathcal{S}(\mathbb{R}^n),$$

where B is the operator in $\mathcal{S}(\mathbb{R}^n)$ given by the Schwartz kernel $\mathcal{B}(x, y) = \mathcal{A}(y, x)$. Therefore $A^T u = Bu \in \mathcal{S}(\mathbb{R}^n)$ for all $u \in \mathcal{S}(\mathbb{R}^n)$. $\qquad \square$

Example 3.6. The δ-function $\delta(x - y)$ (see Example 2.9) can be considered either as a distribution in x depending on the parameter y, or as a distribution in y depending on the parameter x. We have

$$\langle \delta(x - y), f(y) \rangle = f(x), \quad \forall f \in \mathcal{S}(\mathbb{R}^n),$$

that is, $\delta(x - y)$ is the Schwartz kernel of the identity operator.

3.2.   *Oscillatory integrals*

Definition 3.7. We say that a function $a(x, y, \xi)$ on $\mathbb{R}^n_x \times \mathbb{R}^n_y \times \mathbb{R}^n_\xi$ belongs to the class S^m if a is infinitely smooth and

$$|\partial_\xi^\alpha \partial_x^\beta \partial_y^\gamma a(x, y, \xi)| \leq \text{const}_{\alpha, \beta, \gamma} (1 + |\xi|)^{m - |\alpha|}.$$

for all multi-indices α, β, γ. We define $S^{-\infty} := \bigcap_m S^m$, where the intersection is taken over all $m \in \mathbb{R}$.

Obviously, $\xi^{\alpha_1} \partial_\xi^{\alpha_2} \partial_x^\beta \partial_y^\gamma a \in S^{m-|\alpha_2|+|\alpha_1|}$ whenever $a \in S^m$.

Example 3.8. The polynomial $\sum_{|\alpha| \leq m} a_\alpha(x, y)\, \xi^\alpha$ with smooth coefficients a_α belongs to S^m if a_α are bounded with all their derivatives.

Definition 3.9. A function $a(x, y, \xi)$ is said to be *positively homogeneous* of degree m in ξ if $a(x, y, \lambda\xi) = \lambda^m a(x, y, \xi)$ for all $\lambda > 0$.

Example 3.10. Let $a(x, y, \xi)$ be a positively homogeneous of degree m function such that

$$|\partial_x^\beta \partial_y^\gamma a(x, y, \xi)| \leq \text{const}_{\beta,\gamma}, \quad \forall \xi : |\xi| = 1.$$

Then, for every smooth cut-off function $\zeta(\xi)$ vanishing in a neighbourhood of zero and equal to 1 for large ξ, we have $\zeta a \in S^m$.

Definition 3.11. Let m_k be a sequence of real numbers such that $m_k \to -\infty$ as $k \to \infty$, and let $a_{m-k} \in S^{m_k}$. We say that the function $a \in S^m$ admits an asymptotic expansion

$$a(x, y, \xi) \sim \sum_{k=0}^\infty a_{m-k}(x, y, \xi), \quad |\xi| \to \infty, \tag{3.2}$$

if $(a - \sum_{k=0}^l a_{m-k}) \in S^{p_l}$ where $p_l \to -\infty$ as $l \to \infty$ for all $l = 1, 2, \ldots$. We say that a admits the asymptotic expansion (3.2) with a_{m-k} positively homogeneous of degree m_k in ξ if $a \sim \sum_{k=0}^\infty \zeta\, a_{m-k}$, where $\zeta = \zeta(\xi)$ is the same cut-off function as in Example 3.10.

Lemma 3.12. *Let m_k be as in Definition 3.11, and let $m = \max\{m_k\}$. Then for any sequence of functions $a_{m_k} \in S^{m_k}$ there exists a function $a \in S^m$ such that (3.2) holds. This function is determined uniquely modulo $S^{-\infty}$.*

Proof. See [2, Proposition 18.1.3]. □

Definition 3.13. The integral

$$\mathcal{I}_a(x, y) = (2\pi)^{-n} \int e^{i(x-y)\cdot\xi} a(x, y, \xi)\, d\xi \tag{3.3}$$

with $a \in S^m$ is called *oscillatory integral* and the function a is called its *amplitude*.

Remark 3.14. One can replace $(x - y) \cdot \xi$ in (3.3) with a more general *phase function* $\varphi(x, y, \xi)$ which has to be positively homogeneous in ξ of degree 1 and non-degenerate in some appropriate sense (see, for example, [3, 6] or [4]).

One can easily see that for every fixed y the integral (3.3) converges in the sense of distributions in x and, another way round, for every fixed x it converges in the sense of distributions in y. Thus, $\mathcal{I}_a$ can be considered either as a distribution in x depending on the parameter y or as a distribution in y depending on the parameter x. If $m < -n$ then the integral (3.3) is absolutely convergent, so the distribution $\mathcal{I}_a$ coincides with a function. Clearly, this function gets smoother and smoother as $m \to -\infty$; if $a \in S^{-\infty}$ then it is infinitely smooth and bounded with all its derivatives. As a rule, the oscillatory integrals are used for the study of singularities of functions and distributions, and therefore all calculations are carried out modulo $S^{-\infty}$.

3.3. *Pseudodifferential operators*

Lemma 3.15. *Let $a(x, y, \xi) \in S^m$, and let $\sigma(x, \xi)$ be an arbitrary amplitude from S^m such that*

$$\sigma(x, \xi) \sim \sum_\alpha \frac{1}{\alpha!} \, D_\xi^\alpha \partial_y^\alpha a(x, y, \xi)\big|_{y=x}. \tag{3.4}$$

Then $\mathcal{R}(x, y) := \mathcal{I}_a(x, y) - \mathcal{I}_\sigma(x, y)$ is an infinitely differentiable function on $\mathbb{R}_x^n \times \mathbb{R}_y^n$ such that

$$|\partial_x^\beta \partial_y^\gamma \mathcal{R}(x, y)| \leq \mathrm{const}_{\beta, \gamma, N} \, (1 + |x - y|)^{-N} \tag{3.5}$$

for all multi-indices β, γ and positive integers N.

Proof. By Taylor's formula, for all $l = 1, 2, \ldots,$ we have

$$a(x, y, \xi) = \sum_{|\alpha| \leq l} \frac{1}{\alpha!} (y - x)^\alpha \, \partial_y^\alpha a(x, y, \xi)\big|_{y=x} + \sum_{|\alpha|=l+1} (y - x)^\alpha \tilde{a}_\alpha(x, y, \xi),$$

where

$$\tilde{a}_\alpha(x, y, \xi) = \frac{l+1}{\alpha!} \int_0^1 (1 - t)^l \, \partial_z^\alpha a(x, z, \xi)\big|_{z=x+t(y-x)} \, dt.$$

If we substitute this expansion into (3.3), replace $(y - x)^\alpha e^{i(x-y)\cdot\xi}$ with $(-1)^{|\alpha|} D_\xi^\alpha$ and integrate by parts, then we obtain an oscillatory integral with the amplitude

$$\sum_{|\alpha| \le l} \frac{1}{\alpha!} D_\xi^\alpha \partial_y^\alpha a(x, y, \xi)\big|_{y=x} + \sum_{|\alpha|=l+1} D_\xi^\alpha \tilde{a}_\alpha(x, y, \xi)\, d\xi. \tag{3.6}$$

One can easily see that the second sum in (3.6) belongs to S^{m-l-1}.

The above arguments show that, for every σ satisfying (3.4) and every positive integer l, the difference $\mathcal{R}(x, y) = \mathcal{I}_a(x, y) - \mathcal{I}_\sigma(x, y)$ can be represented by the oscillatory integral

$$\int e^{i(x-y)\cdot\xi} b_l(x, y, \xi)\, d\xi$$

with an amplitude $b_l \in S^{m-l-1}$. If $l \ge |\beta| + |\gamma| + m + n$ then

$$\partial_x^\beta \partial_y^\gamma \left((x-y)^\alpha \int e^{i(x-y)\cdot\xi} b_l(x, y, \xi)\, d\xi \right)$$

$$= \partial_x^\beta \partial_y^\gamma \int (D_\xi^\alpha e^{i(x-y)\cdot\xi})\, b_l(x, y, \xi)\, d\xi$$

$$= (-1)^{|\alpha|} \int \partial_x^\beta \partial_y^\gamma \left(e^{i(x-y)\cdot\xi} D_\xi^\alpha b_l(x, y, \xi) \right) d\xi \tag{3.7}$$

and the integral on the right-hand side is absolutely convergent and bounded uniformly with respect to x and y. Since l can be chosen arbitrarily large, it follows that

$$\left| \partial_x^\beta \partial_y^\gamma \left((x-y)^\alpha \mathcal{R}(x, y) \right) \right| \le \mathrm{const}_{\alpha,\beta,\gamma}, \quad \forall \alpha, \beta, \gamma.$$

This implies (3.5). $\qquad\square$

In a similar way one can prove that $\mathcal{I}_a(x, y) - \mathcal{I}_{\sigma'}(x, y)$ is an infinitely smooth function satisfying (3.5) if

$$\sigma' = \sigma'(y, \xi) \sim \sum_\alpha \frac{1}{\alpha!} (-1)^{|\alpha|} D_\xi^\alpha \partial_x^\alpha a(x, y, \xi)\big|_{x=y}. \tag{3.8}$$

Thus, the distribution $\mathcal{I}_a(x, y)$ can be represented, modulo a smooth function satisfying (3.5), by the oscillatory integral with an amplitude independent either of y or of x.

Definition 3.16. We say that an operator A belongs to the class Ψ^m if its Schwartz kernel is given by an oscillatory integral $\mathcal{I}_a(x, y)$ with some amplitude $a \in S^m$. The operator $A \in \Psi^m$ is said to be a *pseudodifferential operator* (PDO) of order m, and the functions σ and σ' satisfying (3.4) and (3.8) are said to be its *symbol* and *dual symbol*, respectively.

Lemma 3.15 and (3.8) imply that

$$\sigma'(y,\xi) \sim \sum_\alpha \frac{1}{\alpha!}(-1)^{|\alpha|} D_\xi^\alpha \partial_y^\alpha \sigma(y,\xi), \quad \sigma(x,\xi) \sim \sum_\alpha \frac{1}{\alpha!} D_\xi^\alpha \partial_x^\alpha \sigma'(x,\xi).$$

$$(3.9)$$

Lemma 3.17. *A PDO $A \in \Psi^m$ continuously maps $\mathcal{S}(\mathbb{R}^n)$ into itself.*

Proof.   According to Definitions 2.20, 3.2 and 3.16, if $A \in \Psi^m$ then

$$Au(x) = (2\pi)^{-n} \int e^{ix \cdot \xi} \left(\int e^{-iy \cdot \xi} a(x,y,\xi)\, u(y)\, dy \right) d\xi$$

with some $a \in S^m$. If $u \in \mathcal{S}(\mathbb{R}^n)$ then the integral with respect to y absolutely converges for each fixed x, ξ and defines a smooth function of (x,ξ) rapidly decreasing with respect to ξ with all its derivatives. The same is true for all integrals obtained by formal differentiation. We have

$$|Au(x)| \leq \int \left| \int e^{-iy \cdot \xi} a(x,y,\xi)\, u(y)\, dy \right| d\xi$$

$$= \int \left| \int \left((1-\Delta_y)^N e^{-iy \cdot \xi}\right)(1+|\xi|^2)^{-N} a(x,y,\xi)\, u(y)\, dy \right| d\xi$$

$$= \int \left| \int e^{-iy \cdot \xi}(1+|\xi|^2)^{-N}(1-\Delta_y)^N \left(a(x,y,\xi)\, u(y)\right) dy \right| d\xi$$

$$\leq \int\!\!\int (1+|\xi|^2)^{-N}|(1-\Delta_y)^N \left(a(x,y,\xi)\, u(y)\right)|\, dy\, d\xi \qquad (3.10)$$

for all positive integers N, where $\Delta_y := \sum_k \partial_{y_k}^2$. The integral on the right-hand side converges and is estimated by a finite linear combination of $\|u\|_{\alpha,\beta}$. Differentiating under the integral sign and integrating by parts, we see that

$$\partial_{x_k} Au(x) = \int e^{ix \cdot \xi} \left(\int e^{-iy \cdot \xi}(i\xi_k)\, a(x,y,\xi)\, u(y)\, dy \right) d\xi$$

$$+ \int e^{ix \cdot \xi} \left(\int e^{-iy \cdot \xi} \partial_{x_k} a(x,y,\xi)\, u(y)\, dy \right) d\xi \qquad (3.11)$$

and

$$x_k \, Au(x) = \int (D_{\xi_k} e^{ix \cdot \xi}) \left(\int e^{-iy \cdot \xi} a(x, y, \xi) \, u(y) \, dy \right) d\xi$$

$$= \int e^{ix \cdot \xi} \left(\int e^{-iy \cdot \xi} a(x, y, \xi) \, y_k \, u(y) \, dy \right) d\xi$$

$$- \int e^{ix \cdot \xi} \left(\int e^{-iy \cdot \xi} \partial_{\xi_k} a(x, y, \xi) \, u(y) \, dy \right) d\xi. \quad (3.12)$$

This implies that $x^\beta \partial_x^\alpha (Au)(x)$ coincides with a finite linear combination of integrals of the same type as $Au(x)$ and therefore is also estimated by a finite linear combination of $\|u\|_{\alpha', \beta'}$. $\qquad \square$

Clearly, $\Psi^m \subset \Psi^l$ whenever $m \leq l$. We shall denote $\Psi^{-\infty} := \bigcap_{m \in \mathbb{R}} \Psi^m$.

Lemma 3.18. *An operator* $R : \mathcal{S}(\mathbb{R}^n) \to \mathcal{S}(\mathbb{R}^n)$ *belongs to* $\Psi^{-\infty}$ *if and only if its Schwartz kernel is an infinitely smooth function satisfying* (3.5).

Exercise 4. Prove Lemma 3.18. *Hint: the Schwartz kernel of an operator from* $\Psi^{-\infty}$ *can be estimated with the use of* (3.7).

By Lemma 3.5, if A is a PDO with an amplitude $a(x, y, \xi)$ then its transposed operator A^T is a PDO with the amplitude $a(y, x, -\xi)$. Therefore, in view of Lemma 2.6, a PDO can be extended to the space $\mathcal{S}'(\mathbb{R}^n)$.

Example 3.19. Every differential operator with smooth coefficients bounded with all their derivatives is a PDO. Indeed, if the Schwartz kernel of A is given by the oscillatory integral with an amplitude $a(x, y, \xi) = \sum_{|\alpha| \leq m} a_\alpha(x, y) \xi^\alpha$ then, by Lemma 2.17,

$$Au(x) = \sum_{|\alpha| \leq m} D_y^\alpha \left(a_\alpha(x, y) \, u(y) \right) |_{y=x}.$$

In particular, if A is the PDO with symbol $\sigma(x, \xi) = \sum_{|\alpha| \leq m} a_\alpha(x) \xi^\alpha$ then

$$Au(x) = \sum_{|\alpha| \leq m} a_\alpha(x) D_x^\alpha u(x),$$

and if A is the PDO with dual symbol $\sigma(y, \xi) = \sum_{|\alpha| \leq m} a_\alpha(y) \xi^\alpha$ then

$$Au(x) = \sum_{|\alpha| \leq m} D_x^\alpha \left(a_\alpha(x) \, u(x) \right).$$

This explains the role of the factor $(2\pi)^{-n}$ appearing in (3.3).

Example 3.20. If A is a PDO with symbol $\sigma_A(x, \xi)$ then, by Lemma 3.15, the symbol σ_{A^T} of the transposed operator A^T admits the asymptotic expansion

$$\sigma_{A^T}(x, \xi) \sim \sum_{\alpha} \frac{1}{\alpha!} D_\xi^\alpha \partial_x^\alpha \sigma(x, -\xi). \tag{3.13}$$

Remark 3.21. In the process of proving Lemmas 3.15, we have shown that $\mathcal{I}_a = \mathcal{I}_b$ where b is the amplitude given by (3.6). Thus two different amplitudes may define the same oscillatory integral. In particular, it may well happen that $\mathcal{I}_a = 0$ but $a \neq 0$, and even $a \notin S^{-\infty}$.

On the other hand, the symbol σ of the PDO A with Schwartz kernel $\mathcal{I}_a$ is defined uniquely modulo $S^{-\infty}$. Indeed, Lemmas 3.15 and 3.18 imply that $\mathcal{I}_a - \mathcal{I}_\sigma \in \Psi^{-\infty}$, so it is sufficient to reconstruct the symbol σ from $\mathcal{I}_\sigma$. Let $\tilde{A}$ be the PDO associated with $\mathcal{I}_\sigma$, so that

$$\tilde{A}u(x) = (2\pi)^{-n} \int e^{i(x-y)\cdot\xi} \sigma(x, \xi) \, u(y) \, dy \, d\xi$$

$$= (2\pi)^{-n/2} \int e^{ix\cdot\xi} \sigma(x, \xi) \, \hat{u}(y) \, d\xi.$$

If u_η is the Schwartz distribution such that $\hat{u}_\eta$ is the delta-function at the point $\eta \in \mathbb{R}^n$ then

$$Au_\eta(x) = \langle u_\eta(y), \mathcal{I}_\sigma(x, y) \rangle = (2\pi)^{-n/2} e^{ix\cdot\xi} \sigma(x, \xi)|_{\xi=\eta}.$$

It follows that $\sigma(x, \eta) = (2\pi)^{-n/2} e^{-ix\cdot\eta} Au_\eta(x)$.

3.4.   Other classes of PDOs

Lemma 3.15 plays the key role in the theory of PDOs. One can consider much more general classes of amplitudes and the corresponding classes of PDOs (see, for example [2]), and usually all classical results remain valid as far as an analogue of Lemma 3.15 holds. For example, given a "weight" functions $g(x, y, \xi)$ and two real numbers $\rho, \delta \in [0, 1]$, we can consider the classes $S_{\rho,\delta}^{m;g}$ which consist of amplitudes satisfying the estimates

$$|\partial_\xi^\alpha \partial_x^\beta \partial_y^\gamma a(x, y, \xi)| \leq \mathrm{const}_{\alpha,\beta,\gamma} \, (g(x, y, \xi))^{m-\rho|\alpha|+\delta|\beta|+\delta|\gamma|}. \tag{3.14}$$

These classes are more convenient than S^m if we want to control not only the smoothness properties of functions but also their behaviour at infinity.

One often has to deal with differential operators depending on an additional parameter h (for instance, the semi-classical parameter). In this case one can introduce a weight function g depending on this parameter and use the classes of amplitudes defined by (3.14) in order to study asymptotics with respect to h (see, for example, [1]).

4. Solution of Partial Differential Equations

Example 4.1. Consider the differential equation $t^2 u'(t) = 0$ on the real line. Obviously, this equation does not have any classical solution apart from $u \equiv$ const. However, if we rewrite this equation as

$$\frac{d}{dt}(t^2 u(t)) - 2t\, u(t) = 0,$$

then we see that any function of the form

$$u(t) = \begin{cases} c_1, & t \geq 0, \\ c_2, & t < 0, \end{cases} \tag{4.1}$$

where c_1, c_2 are constants, is also a solution.

This example shows that if we are looking only for classical solutions then the class of solutions may depend on the way we write down the equation. This problem does not arise if we understand solutions in the sense of distribution. For instance, the derivative of the function (4.1) is the δ-function multiplied by some constant, so $t\, u'(t)$ is equal to 0 as a distribution.

4.1. *Differential equations with constant coefficients*

Let $a(\xi) := \sum_{|\alpha| \leq m} c_\alpha \xi^\alpha$ be a polynomial with constant coefficients c_α and $A = a(D_x) = \sum_{|\alpha| \leq m} c_\alpha D_x^\alpha$ be the differential operator with symbol $a(\xi)$ (see Example 3.19). Then, by Lemma 2.17,

$$Au(x) = \mathcal{F}_{\xi \to x}^{-1} a(\xi) \mathcal{F}_{y \to \xi} u(y), \quad \forall u \in \mathcal{S}'(\mathbb{R}^n).$$

Therefore $Au = f$ if and only if

$$a(\xi)\,\hat{u}(\xi) = \hat{f}(\xi). \tag{4.2}$$

Thus, in order to solve the partial differential equation $Au = f$ with $f \in \mathcal{S}'(\mathbb{R}^n)$ it is sufficient to solve the algebraic equation (4.2).

Example 4.2. If $(a(\xi))^{-1}$ is a polynomially bounded continuous function then the equation $Au = f$ has the unique solution $u(x) = \mathcal{F}_{\xi \to x}^{-1}(a(\xi))^{-1} \hat{f}(\xi) \in \mathcal{S}'(\mathbb{R}^n)$ for every $f \in \mathcal{S}(\mathbb{R}^n)$.

Example 4.3. If $(a(\xi))^{-1}$ is an infinitely differentiable function polynomially bounded with all its derivatives then the equation $Au = f$ has the only solution $u(x) = \mathcal{F}_{\xi \to x}^{-1}(a(\xi))^{-1}f(\xi) \in \mathcal{S}'(\mathbb{R}^n)$ for every $f \in \mathcal{S}'(\mathbb{R}^n)$.

Example 4.4. If $u \in \mathcal{S}'(\mathbb{R}^n)$ and $Au = 0$ then necessarily

$$\operatorname{supp} \hat{u} \subset \Sigma_a := \{\xi : a(\xi) = 0\}.$$

Every distribution $u \in \mathcal{S}'(\mathbb{R}^n)$ whose Fourier transform is given by

$$\langle \hat{u}(\xi), f(\xi) \rangle = \int_{\Sigma_a} v(\xi)\, f(\xi)\, d\Sigma_a(\xi), \quad \forall f \in \mathcal{S}(\mathbb{R}^n), \tag{4.3}$$

where $d\Sigma_a$ is an arbitrary measure and v is an arbitrary integrable function on Σ_a, solves the equation $Au = 0$. If (4.3) holds true and the function v is "sufficiently nice" then

$$u(x) = (2\pi)^{-n/2} \int_{\Sigma_a} e^{ix\xi} v(\xi)\, d\Sigma_a(\xi)$$

is a function on $\mathbb{R}^n$.

A comprehensive exposition of the theory of partial differential equations with constant coefficients can be found in [2, Volume II].

4.2. Non-stationary equations with constant coefficients

If the operator includes the time variable t and we want to solve the Cauchy problem then it is usually more convenient to consider the Fourier transform only with respect to the spatial variables. If, for example,

$$A(\partial_t, D_x) = \partial_t^m + \sum_{|\alpha|=1} c_{m-1,\alpha}\, D_x^\alpha \partial_t^{m-1} + \sum_{|\alpha|=2} c_{m-2,\alpha}\, D_x^\alpha \partial_t^{m-2}$$

$$+ \cdots + \sum_{|\alpha|=m} c_{0,\alpha}\, D_x^\alpha,$$

where $c_{k,\alpha}$ are some constant, then $u(x,t)$ solves the Cauchy problem

$$A(\partial_t, D_x)u(t,x) = 0, \quad \partial_t^k u(t,x)\big|_{t=0} = v_k(x), \quad k = 0, 1, \ldots, m,$$

if and only if $\hat{u}(t,\xi) = \mathcal{F}_{x \to \xi} u(t,x)$ is a solution of the ordinary differential equation

$$A(\partial_t, \xi)\hat{u}(t,\xi) = 0, \quad \partial_t^k \hat{u}(t,\xi)\big|_{t=0} = \hat{v}_k(\xi), \quad k = 0, 1, \ldots, m,$$

where $A(\partial_t, \xi) = \partial_t^m + \sum_{|\alpha|=1} c_{m-1,\alpha}\, \xi^\alpha \partial_t^{m-1} + \cdots + \sum_{|\alpha|=m} c_{0,\alpha}\, \xi^\alpha$. In this case we understand $u(t,x)$ as a family of distributions in x depending on the parameter t, and $\partial_t^k u$ is the family of distributions such that

$$\langle \partial_t^k u(t,x), f(x) \rangle = \partial_t^k \langle u(t,x), f(x) \rangle, \quad \forall f \in \mathcal{S}(\mathbb{R}^n).$$

Example 4.5 (Heat equation). Let $a(x)$ be a semibounded from below polynomial on $\mathbb{R}^n$ and $A = a(D_x)$. Then, for every $v \in \mathcal{S}'(\mathbb{R}^n)$ the distribution

$$u(t,x) = \mathcal{F}_{\xi \to x}^{-1} e^{-ta(\xi)} \hat{v}(\xi)$$

is the only solution of the Cauchy problem

$$\partial_t u + Au = 0, \quad u(0,x) = v(x).$$

If $v \in \mathcal{S}(\mathbb{R}^n)$ then, obviously, $u(t,\cdot) \in \mathcal{S}(\mathbb{R}^n)$ for every t, and the Schwartz kernel of the operator $\exp(-tA) : v(x) \mapsto u(t,x)$ (the so-called heat kernel) is given by the integral

$$(2\pi)^{-n} \int e^{i(x-y)\cdot\xi} e^{-ta(\xi)} \, d\xi$$

which converges in the sense of distributions.

Example 4.6 (Wave equation). If $v \in \mathcal{S}'(\mathbb{R}^n)$ then the distribution

$$u(t,x) = \mathcal{F}_{\xi \to x}^{-1} \cos(t|\xi|) \, \hat{v}(\xi)$$

is the only solution of the Cauchy problem

$$\partial_t^2 u - \Delta u = 0, \quad u(0,x) = v(x), \quad \partial_t u(0,x) = 0,$$

where $\Delta = \sum_k \partial_{x_k}^2$ is the Laplacian. As in the previous example, this implies that $u(t,\cdot) \in \mathcal{S}(\mathbb{R}^n)$ for every t whenever $v \in \mathcal{S}(\mathbb{R}^n)$.

4.3. Elliptic (pseudo)differential equations

Lemma 4.7 (Composition of PDOs). *If $A \in \Psi^{m_1}$ and $B \in \Psi^{m_2}$ then $AB \in \Psi^{m_1+m_2}$ and the symbol σ_{AB} of the PDO AB is given by the asymptotic series*

$$\sigma_{AB}(x,\xi) \sim \sum_\alpha \frac{1}{\alpha!} D_\xi^\alpha \sigma_A(x,\xi) \, \partial_x^\alpha \sigma_B(x,\xi), \tag{4.4}$$

where σ_A and σ_B are the symbols of A and B respectively.

Proof. If A and B are given by the oscillatory integrals (3.3) with $\sigma_A(x, \xi)$ and $\sigma'_B(y, \xi)$ respectively, then

$$Av(x) = (2\pi)^{-n} \int e^{i(x-y)\cdot\xi} \sigma_A(x, \xi)\, v(y)\, dy\, d\xi$$

$$= (2\pi)^{-n/2} \int e^{ix\cdot\xi} \sigma_A(x, \xi)\, \hat{v}(\xi)\, d\xi$$

and

$$Bu(x) = (2\pi)^{-n} \int e^{i(x-y)\cdot\xi} \sigma'_B(y, \xi)\, u(y)\, dy\, d\xi$$

$$= \mathcal{F}^{-1}_{\xi \to x} \left((2\pi)^{-n/2} \int e^{-iy\cdot\xi} \sigma'_B(y, \xi)\, u(y)\, dy \right).$$

Therefore

$$ABu(x) = (2\pi)^{-n} \int e^{ix\cdot\xi} \sigma_A(x, \xi) \left(\int e^{-iy\cdot\xi} \sigma'_B(y, \xi)\, u(y)\, dy \right) d\xi,$$

that is, the Schwartz kernel of AB coincides with the oscillatory integral with the amplitude $\sigma_A(x, \xi)\, \sigma'_B(y, \xi)$. By (3.4), we have

$$\sigma_{AB}(x, \xi) \sim \sum_\alpha \frac{1}{\alpha!} D^\alpha_\xi \left(\sigma_A(x, \xi)\, \partial^\alpha_x \sigma'_B(x, \xi) \right).$$

Now (4.4) is obtained by substituting the first expansion (3.9) and rearranging terms in the asymptotic series. $\qquad\square$

Definition 4.8. A PDO $A \in \Psi^m$ is said to be *classical* if its symbol admits an asymptotic expansion into the series (3.2) with a_{m-k} positively homogeneous in ξ of degree $m - k$. The spaces of classical PDOs $A \in \Psi^m$ and their symbols are denoted by Ψ^m_{cl} and S^m_{cl}, respectively.

Obviously, every differential operator is a classical PDO.

Definition 4.9. If $A \in \Psi^m_{\mathrm{cl}}$ then the leading homogeneous term a_m in the expansion of σ_A is said to be the *principal symbol* of the operator A. The operator A is said to be elliptic if $a_m(x, \xi) \neq 0$ whenever $\xi \neq 0$.

Definition 4.10. The operator B is said to be a *left parametrix* of A if $BA - I \in \Psi^{-\infty}$.

If B is a left parametrix of A and $Au = f$ then $(I + R)u = Bf$ where $R \in \Psi^{-\infty}$. The "remainder" operator R often turns out to be compact in a suitable function space H. In this case the existence of a parametrix implies that the subspace $\{u \in H : Au = 0\}$ is of finite dimension, and that the equation $Au = f$ has a solution for all f from a subspace of finite codimension.

Theorem 4.11. *Every elliptic classical PDO $A \in \Psi_{\mathrm{cl}}^m$ has a left parametrix $B \in \Psi_{\mathrm{cl}}^{-m}$.*

Proof. We shall construct the symbol of B as an asymptotic series of positively homogeneous in ξ functions $b_{-m-k}(x, \xi)$ of degree $-m - k$. If we substitute the formal series $\sigma_A = \sum_{k=0}^m a_{m-k}$ and $\sigma_B = \sum_{k=0}^\infty b_{-m-k}$ into (4.4), collect together homogeneous terms of the same degree, equate the first term to 1 and others to zero, then we obtain a recurrent system of differential equations of the form

$$a_m b_{-m} = 1,$$

$$a_m b_{-m-1} = L_1(a_m, a_{m-1}, b_{-m})$$

$$\vdots$$

$$a_m b_{-m-k} = L_k(a_m, a_{m-1} \ldots, a_{m-k}, b_{-m}, b_{-m-1}, \ldots, b_{-m-k+1})$$

$$\vdots$$

where $L_k(a_m, a_{m-1} \ldots, a_{m-k}, b_{-m}, b_{-m-1}, \ldots, b_{-m-k+1})$ are some polynomials of the functions $a_m, a_{m-1} \ldots, a_{m-k}, b_{-m}, b_{-m-1}, \ldots, b_{-m-k+1}$ and their derivatives. If $b_{-m} = a_m^{-1}$,

$$b_{-m-k} = a_m^{-1} L_k(a_m, \ldots, a_{m-k}, b_{-m}, \ldots, b_{-m-k+1}), \quad k = 1, 2, \ldots,$$

and $\sigma_B \sim \sum_{k=0}^\infty b_{-m-k}$ then $\sigma_{AB} = 1$ modulo $S^{-\infty}$, that is, $AB - I \in \Psi^{-\infty}$. $\quad\square$

Exercise 5. Prove that

(1) $A \in \Psi_{\mathrm{cl}}^m$ if and only if $A^T \in \Psi_{\mathrm{cl}}^m$, where A^T is the transposed operator (see Definition 3.3);
(2) A is elliptic if and only if A^T is elliptic;
(3) if $A \in \Psi_{\mathrm{cl}}^m$ is elliptic then there exists an operator $B \in \Psi_{\mathrm{cl}}^{-m}$, called a *right parametrix* of A, such that $AB - I \in \Psi^{-\infty}$.

Hint: deduce (3) from (1), (2) and Theorem 4.11.

Remark 4.12. Let $A \in \Psi_{\mathrm{cl}}^m$ be a classical PDO and $\mathcal{O} \subset \mathbb{R}_x^n \times (\mathbb{R}_\xi^n \setminus \{0\})$ be a conic with respect to ξ subset (the word conic means that $(x, \lambda \xi) \in \mathcal{O}$ for all $\lambda > 0$ whenever $(x, \xi) \in \mathcal{O}$). If the principal symbol of A is separated from 0 on the set $\mathcal{O} \cap \{|\xi| = 1\}$ then, exactly in the same way, one can construct a PDO $B \in \Psi_{\mathrm{cl}}^{-m}$ such that $\sigma_{BA} = 1$ on $\mathcal{O}$. Such an operator is called a *microlocal parametrix* of A in $\mathcal{O}$.

4.4. General partial differential equations with variable coefficients

An arbitrary partial differential operator does not necessarily have a pseudodifferential parametrix. However, quite often one can construct a parametrix in the form of a general oscillatory integral (Remark 3.14) or a PDO which belongs to a more general class (Subsection 3.4). The procedure remains almost the same as in the proof Theorem 4.11: we formally replace the amplitude with an asymptotic series, substitute the integral into the equation, get rid of the variable y in the new amplitude (like we did in the proof of Lemma 3.15), collect together the terms of the same order, equate them to zero and try to solve these equations. Note that in the general case the equations may also involve the unknown phase function, and that the terms in the asymptotic expansions may not be homogeneous in ξ (in which case the words "terms of the same order" simply mean that these terms satisfy estimates of the form (3.14) with the same m, ρ and δ).

5. Singularities of Functions and Distributions

5.1. What is microlocal analysis

Suppose that we want to describe singularities of a function $f(x)$ on $\mathbb{R}^n$. In classical analysis one only deals with the variables x, and the typical statements look like "*the function f has a singularity at the point x_0*" or "*f is smooth in a neighbourhood of x_0*". However, the function may well be smooth in one direction and non-smooth in another direction, so such statements contain a limited information. More detailed description of singularities should involve additional variables ξ specifying the directions in which the function is not smooth. In other words, the set of singularities should be a subset of $\mathbb{R}_x^n \times \mathbb{R}_\xi^n$, and then we say that "*f is not smooth at the point $(x, \xi) \in \mathbb{R}_x^n \times \mathbb{R}_\xi^n$*" if f is not smooth at the point x in the direction ξ. This is the main idea of microlocal analysis; the word "microlocal" simply means that we conduct analysis of functions in the space $\mathbb{R}^n_x \times \mathbb{R}^n_\xi$ of dimension $2n$, even though the functions themselves are defined on the n-dimensional space.

5.2. *Singular supports and wave front sets*

Definition 5.1. If $u \in \mathcal{S}'(\mathbb{R}^n)$ then the *singular support* of u is defined by

$$\text{sing supp}\, u := \mathbb{R}^n \setminus \Omega_u,$$

where Ω_u is the maximal open subset of $\mathbb{R}^n$ such that $u|_{\Omega_u} \in C^\infty(\Omega_u)$.

Definition 5.2. If $u \in \mathcal{S}'(\mathbb{R}^n)$ then the *wave front set* of u is defined by

$$\text{WF}\, u := (\mathbb{R}^n_x \times \mathbb{R}^n_\xi) \setminus \mathcal{O}_u,$$

where $\mathcal{O}_u$ is the maximal open subset of $\mathbb{R}^n_x \times \mathbb{R}^n_\xi$ with following property: for every point $(x_0, \xi_0) \in \mathcal{O}_u$ there exists a cut-off function $\chi \in C_0^\infty(\mathbb{R}^n)$ equal to 1 in a neighbourhood of x_0 and a conic neighbourhood Ω_{ξ_0} of ξ_0 such that the Fourier transform $\mathcal{F}_{x\to\xi}(\chi(x)u(x))$ vanishes in Ω_{ξ_0} faster than any power of $|\xi|$ as $|\xi| \to \infty$.

Note that a compactly supported distribution u coincides with an infinitely smooth function if and only if its Fourier transform $\hat{u}(\xi)$ vanishes faster than any power of $|\xi|$ as $|\xi| \to \infty$. Indeed, if u is infinitely differentiable then $u \in \mathcal{S}(\mathbb{R}^n)$, and if $\hat{u}(\xi)$ vanishes faster than any power of $|\xi|$ as $|\xi| \to \infty$ then

$$u(x) = (2\pi)^{-n/2} \int e^{ix\cdot\xi}\, \hat{u}(\xi)\, d\xi$$

and we can differentiate under the integral sign infinitely many times. This implies that the projection of $\text{WF}\, u$ onto $\mathbb{R}^n_x$ coincides with $\text{sing supp}\, u$.

Definition 5.2 can be rewritten as follows.

Definition 5.3. The point (x_0, ξ_0) does not belong to $\text{WF}\, u$ if there exists a C_0^∞-function χ equal to 1 in a neighbourhood of x_0 and a conic neighbourhood $\mathcal{O}_{\xi_0}$ such that for every classical PDO $Q_{a,\chi}$ with dual symbol $a(\xi)\chi(y)$ we have $Q_{a,\chi}u \in C^\infty(\mathbb{R}^n)$ whenever $\text{supp}\, a \in \mathcal{O}_{\xi_0}$.

Lemma 5.4. Let $R \in \Psi^{-\infty}$ and $\mathcal{R}(x,y)$ be the Schwartz kernel of R. Then

$$Ru(x) = \langle u(y), \mathcal{R}(x,y)\rangle, \quad \forall u \in \mathcal{S}'(\mathbb{R}^n). \tag{5.1}$$

Note that, in view of Lemma 3.18, we have $\mathcal{R}(x, \cdot) \in \mathcal{S}(\mathbb{R}^n)$ for each fixed $x \in \mathbb{R}^n$. Therefore the expression on the right-hand side of (5.1) makes sense.

Proof. If $u \in \mathcal{E}'(\mathbb{R}^n)$ then, by Theorem 2.15 and Lemma 3.5, we have

$$\langle Ru, f \rangle = \langle u, R^T f \rangle = \sum_{|\alpha| \leq m} (-1)^{|\alpha|} \langle u_\alpha(y), \partial_y^\alpha (R^T f)(y) \rangle$$

$$= \sum_{|\alpha| \leq m} (-1)^{|\alpha|} \int u_\alpha(y) \left(\partial_y^\alpha \int \mathcal{R}(x, y) f(x) \, dx \right) dy$$

$$= \sum_{|\alpha| \leq m} (-1)^{|\alpha|} \left(\int u_\alpha(y) \partial_y^\alpha \mathcal{R}(x, y) \, dy \right) f(x) \, dx$$

$$= \int \langle u(y), \mathcal{R}(x, y) \rangle f(x) \, dx$$

for all $f \in \mathcal{S}(\mathbb{R}^n)$, where u_α are some continuous compactly supported functions. Therefore (5.1) holds true whenever $u \in \mathcal{E}'(\mathbb{R}^n)$.

If $u \notin \mathcal{E}'(\mathbb{R}^n)$ then we choose a function $\chi \in C_0^\infty(\mathbb{R}^n)$ which is equal to 1 in a neighbourhood of origin and consider the family of distributions $u_t(x) := \chi(tx) u(x)$. If $\chi \equiv 0$ outside the ball $\{|x| \leq R\}$ then, applying Taylor's formula to $\partial_x^\alpha \chi$, we obtain

$$\sup_y |y^\beta \partial_x^\alpha (\chi(tx) - \chi(0))| \leq R^{|\beta|+1} t^{|\alpha|+1} \sum_{|\alpha'|=|\alpha|+1} \sup_x |\partial_x^{\alpha'} \chi(x)|$$

for all multi-indices α, β. This implies that $\chi(tx) f(x) \xrightarrow{S} f(x)$ and, therefore, $\langle u_t, f \rangle \to \langle u, f \rangle$ for all $f \in \mathcal{S}(\mathbb{R}^n)$ as $t \to 0$. Now the lemma is proved by applying (5.1) to the distribution $u_t \in \mathcal{E}'(\mathbb{R}^n)$ for each fixed t and letting $t \to 0$ in the identity $\langle u_t, R^T f \rangle = \int \langle u_t(y), \mathcal{R}(x, y) \rangle f(x) \, dx$. □

Corollary 5.5. *If $R \in \Psi^{-\infty}$ then $Ru \in C^\infty(\mathbb{R}^n)$ whenever $u \in \mathcal{S}'(\mathbb{R}^n)$ and $Ru \in \mathcal{S}(\mathbb{R}^n)$ whenever $u \in \mathcal{E}'(\mathbb{R}^n)$.*

Proof. By Taylor's formula

$$\varepsilon^{-1} \left(\mathcal{R}(x_1, \ldots, x_k + \varepsilon, \ldots, x_n, y) - \mathcal{R}(x, y) \right) - \partial_{x_k} \mathcal{R}(x, y)$$

$$= \varepsilon \int_0^1 \partial_{x_k}^2 \mathcal{R}(x_1, \ldots, x_k + t\varepsilon, \ldots, x_n, y) (1 - t) \, dt.$$

If the Schwartz kernel $\mathcal{R}(x, y)$ satisfies (3.5) then the above identity implies that

$$\varepsilon^{-1} \left(\mathcal{R}(x_1, \ldots, x_k + \varepsilon, \ldots x_n, y) - \mathcal{R}(x, y) \right) \xrightarrow{S} \partial_{x_k} \mathcal{R}(x, y), \quad \varepsilon \to 0,$$

for every fixed $x = (x_1, \ldots, x_n) \in \mathbb{R}^n$ and all $k = 1, 2, \ldots, n$. Therefore, for every $u \in \mathcal{S}'(\mathbb{R}^n)$, the function $\langle u(y), \mathcal{R}(x, y) \rangle$ of variable x is differentiable and

$$\partial_{x_k} \langle u(y), \mathcal{R}(x, y) \rangle = \langle u(y), \partial_{x_k} \mathcal{R}(x, y) \rangle$$

for all $k = 1, 2, \ldots, n$. Since the derivatives of $\mathcal{R}$ also satisfy (3.5), the function $\langle u(y), \mathcal{R}(x, y) \rangle$ is infinitely differentiable and $\partial_x^\alpha \langle u(y), \mathcal{R}(x, y) \rangle = \langle u(y), \partial_x^\alpha \mathcal{R}(x, y) \rangle$ for all multi-indices α. If, in addition, $u \in \mathcal{E}'(\mathbb{R}^n)$ then, applying Theorem 2.15, one can easily show that this function and all its derivatives vanish faster than any power of $|x|$ as $|x| \to \infty$. $\qquad \square$

Lemma 5.6. *The point (x_0, ξ_0) does not belong to* WF u *if and only if there exists a classical PDO Q such that $Qu \in \mathcal{S}(\mathbb{R}^n)$ and the principal symbol of Q does not vanish at (x_0, ξ_0).*

Proof. If $(x_0, \xi_0) \notin WFu$ then we can take $Q = \tilde{\chi} Q_{a,\chi}$, where $Q_{a,\chi}$ is a classical PDO from Definition 5.3 and $\tilde{\chi}$ is an arbitrary C_0^∞-function equal to 1 in a neighbourhood of x_0

Conversely, assume that $Qu \in \mathcal{S}(\mathbb{R}^n)$ for some classical PDO $Q \in \Psi_{cl}^m$ whose principal symbol does not vanish at (x_0, ξ_0). There exist a neighbourhood Ω_{x_0} and a conic neighbourhood Ω_{ξ_0} such that the principal symbol of Q is separated from zero on $\mathcal{O} \cap \{|\xi| = 1\}$, where $\mathcal{O} := \Omega_{x_0} \times \Omega_{\xi_0}$. If P is a microlocal parametrix of Q in $\mathcal{O}$ (see Remark 4.12) then, by Lemma 4.7, we have $Q_{a,\chi} P Q - Q_{a,\chi} \in \Psi^{-\infty}$ whenever $\mathrm{supp}\,\chi \subset \Omega_{x_0}$ and $\mathrm{supp}\,a \subset \mathcal{O}_{\xi_0}$. In view of Corollary 5.5, this implies that $Q_{a,\chi} u \in C^\infty(\mathbb{R}^n)$ for all such functions χ and a. $\qquad \square$

Corollary 5.7. *If P is a classical PDO whose symbol vanishes in a conic neighbourhood $\mathcal{O}$ of (x_0, ξ_0) then $(x_0, \xi_0) \notin$ WF (Pu).*

Proof. By Lemma 4.7, we have $QP \in \Psi^{-\infty}$ for every classical PDO Q whose symbol vanishes outside a smaller conic subset $\tilde{\mathcal{O}} \subset \mathcal{O}$. In view of Corollary 5.5, this implies that $Q(Pu) \in C^\infty(\mathbb{R}^n)$ for every such a PDO Q. $\qquad \square$

Corollary 5.8. *Let A be a classical PDO whose principal symbol does not vanish at (x_0, ξ_0). Then $(x_0, \xi_0) \in$ WF (Au) if and only if $(x_0, \xi_0) \in$ WF u.*

Proof. If $(x_0, \xi_0) \notin$ WF (Au) then there exists a classical PDO Q whose principal symbol does not vanish at (x_0, ξ_0) and such that $QAu \in \mathcal{S}(\mathbb{R}^n)$. Then, since the principal symbol of QA does not vanish at (x_0, ξ_0), we have $(x_0, \xi_0) \notin$ WF u.

Let us now assume that $(x_0, \xi_0) \in \mathrm{WF}\,(Au)$ and let B be a microlocal parametrix of A in a neighbourhood of (x_0, ξ_0). Then, applying Corollary 5.7 with $P = BA - I$, we see that $(x_0, \xi_0) \notin \mathrm{WF}\,(BAu - u)$. At the same time, by the above, $(x_0, \xi_0) \in \mathrm{WF}\,(BAu)$. This implies that $(x_0, \xi_0) \in \mathrm{WF}\,u$.

$\square$

Lemma 5.6 and Corollaries 5.7, 5.8 show that PDOs play the same role in microlocal analysis as smooth cut-off functions in classical analysis. One can construct, for example, a microlocal partition of unity using the classical PDOs and represent an arbitrary function as the sum of functions with small wave front sets (this procedure is called *microlocalization*).

5.3.   *Propagation of singularities*

If A is an elliptic PDO then, by Corollary 5.8, we have $\mathrm{WF}\,(Au) = \mathrm{WF}\,u$ and, consequently, $\mathrm{sing\,supp}\,(Au) = \mathrm{sing\,supp}\,u$. In other words, a solution u of an elliptic (pseudo)differential equation $Au = f$ has the same singularities as the function f. This is usually not true if the operator A is not elliptic: the solutions u of a non-elliptic equation $Au = f$ may have additional singularities. This effect is called propagation of singularities.

Propagation of singularities for solutions of general partial differential equations is described by the following theorem (see, for example, [3]).

Theorem 5.9. *Let B be a classical PDO with a real principal symbol $b_m(x, \xi)$ and let $(x(t), \xi(t))$ be a solution of the Hamiltonian system*

$$\dot{x}(t) = \partial_\xi b_m(x(t), \xi(t)), \qquad \dot{\xi}(t) = -\partial_x b_m(x(t), \xi(t)).$$

Assume that $b_m(x(t), \xi(t)) = 0$ and $(x(t), \xi(t)) \notin \mathrm{WF}\,(Bu)$ for all $t \in (t_1, t_2)$. Then either $(x(t), \xi(t)) \notin \mathrm{WF}\,u$ or $(x(t), \xi(t)) \in \mathrm{WF}\,u$ for all $t \in (t_1, t_2)$.

Remark 5.10. One can easily see that $b_m(x(t), \xi(t))$ is constant, so $b_m(x(t), \xi(t)) = 0$ for all $t \in (t_1, t_2)$ provided that $b_m(x(t_0), \xi(t_0)) = 0$ for some fixed $t_0 \in (t_1, t_2)$.

6.   Solutions to Exercises

Solution 1. If $\varepsilon < 1$ then

$$\int_{|t| > \varepsilon} t^{-1} f(t)\, dt = \int_{|t| \geq 1} t^{-1} f(t)\, dt + \int_{1 > |t| > \varepsilon} t^{-1} f(t)\, dt.$$

The first integral is estimated by

$$\int_{|t|\geq 1} t^{-2}\,|t\,f(t)|\,dt \leq 2\sup_{s\in\mathbb{R}}|s\,f(s)|\int_1^\infty t^{-2}\,dt = 2\sup_{s\in\mathbb{R}}|s\,f(s)|.$$

Since $|f(t) - f(-t)| \leq 2t\sup_{s\in[-t,t]}|f'(s)|$, the second integral is estimated as follows

$$\left|\int_{1>|t|>\varepsilon} t^{-1}\,f(t)\,dt\right| = \left|\int_\varepsilon^1 t^{-1}\,(f(t) - f(-t))\,dt\right| \leq 2\sup_{s\in\mathbb{R}}|f'(s)|.$$

Thus $|\int_{|t|>\varepsilon} t^{-1}\,f(t)\,dt| \leq 2\sup_{s\in\mathbb{R}}|s\,f(s)| + 2\sup_{s\in\mathbb{R}}|f'(s)|$ for all $\varepsilon \in (0,1]$ and $f \in \mathcal{S}(\mathbb{R})$. This implies that $u_0 : f \mapsto \lim_{\varepsilon\to 0}\int_{|t|>\varepsilon} t^{-1}\,f(t)\,dt$ is a continuous functional on $\mathcal{S}(\mathbb{R})$, that is, $u_0 \in \mathcal{S}'(\mathbb{R})$.

Solution 2. Clearly, the distribution u_0 from Exercise 2 satisfies this condition. If u is another distribution with the same property then $\langle u - u_0, f\rangle$ depends only on the value of f at the origin. The map $f(0) \to \langle u - u_0, f\rangle$ is a linear functional on $\mathbb{C}$, and therefore $\langle u - u_0, f\rangle = cf(0)$ with some constant c. This implies that $u = u_0 + c\delta$, where δ is the δ-function at the origin. Conversely, every distribution $u = u_0 + c\delta$ has the required property.

Solution 3. Let $c_k = u(a_k + 0) - u(a_k - 0)$ be the jumps of u at the points a_k, and let

$$v(x) = \begin{cases} u'(x), & x \in (a_k, a_{k+1}), \quad k = 1, \ldots, m-1, \\ 0, & x < a_1 \text{ or } x > a_m. \end{cases}$$

Integrating by parts we obtain $-\langle u, f'\rangle = -\int u f'\,dx = \int v f\,dx + \sum_{k=1}^m c_k f(a_k)$. Therefore $u' = v + \sum_{k=1}^m c_k \delta_{a_k}$ where δ_{a_k} are the δ-functions at the points a_k.

Solution 4 (proof of Lemma 3.18). Let $\mathcal{R}(x,y)$ be the Schwartz kernel of R. If $R \in \Psi^{-\infty}$ then $\mathcal{R}(x,y)$ is infinitely smooth (since we can differentiate under the integral sign), and the required estimates follow from (3.7).

Conversely, if $\mathcal{R}$ is smooth and satisfies (3.5) then it is represented by the oscillatory integral with the amplitude $a(x,\xi) = (2\pi)^{n/2}\mathcal{F}_{z\to\xi}\tilde{\mathcal{R}}(x,z)$, where $\tilde{\mathcal{R}}$ is defined by the equality $\tilde{\mathcal{R}}(x, x-y) = \mathcal{R}(x,y)$.

Solution 5. Let $A \in \Psi^m$, and let σ be the symbol of A, so that

$$\mathcal{A}(x,y) = (2\pi)^{-n}\int e^{i(x-y)\cdot\xi}\sigma(x,\xi)\,d\xi$$

modulo a function $\mathcal{R}(x,y)$ defining an operator from $\Psi^{-\infty}$ (see Lemma 3.18). Lemma 3.5 implies that A^T is a PDO whose dual symbol coincides with σ. Now (1) and (2) follow from (3.9).

By (2), A is elliptic if and only if A^T is elliptic. Applying Theorem 4.11, let us find a PDO $B_1 \in \Psi_{cl}^{-m}$ such that $B_1 A^T - I \in \Psi^{-\infty}$. In view of Lemmas 3.5 and 3.18, we have $A B_1^T - I = (B_1 A^T - I)^T \in \Psi^{-\infty}$. It remains to notice that, by (1), $B_1^T \in \Psi_{cl}^{-m}$.

References

[1] M. Dimassi and J. Sjöstrand, *Spectral Asymptotics in the Semi-classical Limit.* LMS Lecture Notes Series, Vol. 268. Cambridge University Press (1999).

[2] L. Hörmander, *The Analysis of Linear Partial Differential Operators, I–IV.* Springer-Verlag, New York (1984).

[3] M. Shubin, *Pseudodifferential Operators and Spectral Theory.* Nauka, Moscow (1978). English transl. Springer-Verlag (1987).

[4] M. Taylor, *Pseudodifferential Operators.* Princeton University Press, Princeton, NJ (1981).

[5] F. Treves, *Introduction to Pseudodifferential and Fourier Integral Operators,* I, II. Plenum Press, New York (1980).

[6] Y. Safarov and D. Vassiliev, *The Asymptotic Distribution of Eigenvalues of Partial Differential Operators.* American Mathematical Society, Providence, RI (1996).

Chapter 3

C*-algebras

Cho-Ho Chu

School of Mathematical Sciences
Queen Mary University of London, London E1 4NS, UK
c.chu@qmul.ac.uk

C*-algebras are an important area of research in functional analysis and provide a natural framework for *non-commutative geometry*. In physics, they are used to model a physical system in quantum mechanics. This chapter is based on the ten lectures on *C*-algebras* first given in 2008 at the London Taught Course Centre. It is intended to be a *brief* introduction to the basic theory of C*-algebras and their representations on Hilbert spaces, as well as the Murray–von Neumann classification of von Neumann algebras. The only prerequisite for what follows is a basic knowledge of algebra, topology and functional analysis at an advanced undergraduate level. A selection of books for further reading is listed at the end of the chapter.

1. Abstract and Concrete C*-algebras

All vector spaces $\mathcal{A}$ throughout are over the complex number field $\mathbb{C}$ so that a *conjugate* linear map $x \in \mathcal{A} \mapsto x^* \in \mathcal{A}$ is called an *involution* if it has period 2, that is, $(x^*)^* = x$. The identity of an algebra will always be denoted by $\mathbf{1}$. We recall that a Banach space is a complete normed vector space and refer to [15] for a review of basic concepts and theorems in functional analysis.

Definition 1.1. A Banach space $(\mathcal{A}, \| \cdot \|)$ is called a *Banach algebra* if it is an algebra in which the multiplication satisfies

$$\|xy\| \leq \|x\|\|y\| \qquad (x, y \in \mathcal{A}).$$

We note that multiplication in a Banach algebra is continuous. A Banach algebra $\mathcal{A}$ is called *unital* if it contains an identity, in which case there is an equivalent norm $|\cdot|$ on $\mathcal{A}$ such that $(\mathcal{A}, |\cdot|)$ is a Banach algebra and $|\mathbf{1}| = 1$, where

$$|x| := \sup\{\|xy\| : \|y\| \leq 1\}.$$

Therefore, there is no loss of generality to assume, in the sequel, that in a unital Banach algebra we always have $\|\mathbf{1}\| = 1$. A *Banach $*$-algebra* is a Banach algebra $\mathcal{A}$ which admits an involution $* : \mathcal{A} \longrightarrow \mathcal{A}$ satisfying $(xy)^* = y^*x^*$ and $\|x^*\| = \|x\|$ for all $x, y \in \mathcal{A}$.

Definition 1.2. A Banach $*$-algebra $\mathcal{A}$ is called a *C^*-algebra* if its norm and involution satisfy

$$\|x^*x\| = \|x\|^2 \qquad (x \in \mathcal{A}).$$

This is the definition for an '*abstract*' C*-algebra.

   In the sequel, $\mathcal{A}$ always denotes a C*-algebra unless otherwise stated.

Remark 1.3. A Banach algebra $\mathcal{A}$ which admits an involution $*$ satisfying $(xy)^* = y^*x^*$ and $\|x^*x\| = \|x\|^2$ is a C*-algebra since $\|x\| = \|x^*\|$ follows from $\|x\|^2 = \|x^*x\| \leq \|x^*\|\|x\|$ whence $\|x\| \leq \|x^*\|$ and the inequality can be reversed for x^*. Historically C*-algebras were first defined with the extra condition that $1 + x^*x$ is invertible, but this was later found superfluous.

Example 1.4. Let Ω be a locally compact Hausdorff space and let $C_0(\Omega)$ be the Banach algebra of complex continuous functions on Ω vanishing at infinity, with pointwise multiplication and the supremum norm

$$\|f\| = \sup\{|f(\omega)| : \omega \in \Omega\} \qquad (f \in C_0(\Omega)).$$

It is a Banach $*$-algebra with the complex conjugation as involution

$$f^*(\omega) = \overline{f(\omega)} \qquad (\omega \in \Omega).$$

If Ω is compact, then $C_0(\Omega)$ coincides with the algebra $C(\Omega)$ of all complex continuous functions on Ω.

   Further, $C_0(\Omega)$ is a C*-algebra which is also abelian. In fact, all abelian C*-algebras are of this form.

   A map $\varphi : \mathcal{A} \longrightarrow \mathcal{B}$ between two C*-algebras is called a $*$-*map* if it preserves the involution: $\varphi(a^*) = \varphi(a)^*$. As usual, we denote by $\mathcal{A}^*$ the *dual* *space* of a Banach space $\mathcal{A}$, consisting of all continuous linear functionals on $\mathcal{A}$.

There are two fundamental theorems in the theory of C*-algebras, both bear the name of *Gelfand–Naimark theorem*. The first one characterises abelian C*-algebras, and the second one exhibits a concrete representation of a C*-algebra as an algebra of operators on Hilbert spaces. We outline the proof of these two theorems below, leaving out some details. To complete these details and the proof, one needs to use several results developed in the following sections.

Theorem 1.5 (Gelfand–Naimark Theorem). *Let $\mathcal{A}$ be an abelian C*-algebra. Then it is isometrically $*$-isomorphic to the C*-algebra $C_0(\Omega_{\mathcal{A}})$ of complex continuous functions on a locally compact Hausdorff space $\Omega_{\mathcal{A}}$, vanishing at infinity.*

Proof. We equip the dual space $\mathcal{A}^*$ of $\mathcal{A}$ with the weak* topology in which a net (f_α) in $\mathcal{A}^*$ weak* converges to $f \in \mathcal{A}^*$ if and only if the net $(f_\alpha(a))$ in $\mathbb{C}$ converges to the number $f(a)$ for all $a \in \mathcal{A}$.

Each algebra homomorphism $\omega : \mathcal{A} \longrightarrow \mathbb{C}$, that is, a multiplicative linear functional ω, satisfies $\|\omega\| \leq 1$. We sometimes call ω a *character* if $\omega \neq 0$. Let

$$\Omega_{\mathcal{A}} = \{\omega \in \mathcal{A}^* : \omega \text{ is a non-zero homomorphism from } \mathcal{A} \text{ to } \mathbb{C}\}$$

and let $\Omega = \Omega_{\mathcal{A}} \cup \{0\}$.

Then Ω is weak* closed in the closed unit ball $\mathcal{A}_1^* = \{f \in \mathcal{A}^* : \|f\| \leq 1\}$. By Tychonoff's theorem, $\mathcal{A}_1^*$ is compact in the weak* topology of $\mathcal{A}^*$. Hence Ω is weak* compact. Since $\{0\}$ is closed, $\Omega_{\mathcal{A}}$ is open in Ω and is therefore locally compact.

We define a map $\widehat{\ } : \mathcal{A} \longrightarrow C_0(\Omega_{\mathcal{A}})$, called the *Gelfand transform*, by

$$\widehat{a}(\omega) = \omega(a) \qquad (\omega \in \Omega).$$

This map is well defined since $\widehat{a}$ is clearly continuous on $\Omega_{\mathcal{A}}$ and given $\varepsilon > 0$, the set

$$\{\omega \in \Omega_{\mathcal{A}} : |\widehat{a}(\omega)| \geq \varepsilon\}$$

is closed in Ω and hence weak* compact.

It follows that the Gelfand transform $\widehat{\ }$ is an isometric algebra isomorphism preserving the involution, that is, $\widehat{a^*} = \overline{\widehat{a}}$. $\qquad\square$

Definition 1.6. We call $\Omega_\mathcal{A}$, defined in the proof of Theorem 1.5, the *spectrum* of $\mathcal{A}$.

Example 1.7. Let $\{\mathcal{A}_\alpha : \alpha \in J\}$ be a family of C*-algebras. The Cartesian product

$$\underset{\alpha \in J}{\times} \mathcal{A}_\alpha$$

is clearly a *-algebra in the pointwise product and involution. We define the C*-*direct sum* of $\{\mathcal{A}_\alpha\}_{\alpha \in J}$ to be the following *-subalgebra of $\times_{\alpha \in J} \mathcal{A}_\alpha$:

$$\bigoplus_{\alpha \in J} \mathcal{A}_\alpha = \left\{ (a_\alpha) \in \underset{\alpha \in J}{\times} \mathcal{A}_\alpha : \sup_\alpha \|a_\alpha\| < \infty \right\}.$$

Then $\bigoplus_{\alpha \in J} \mathcal{A}_\alpha$ is a C*-algebra with the norm $\|(a_\alpha)\| = \sup_\alpha \|a_\alpha\|$.

Example 1.8. Let H be a Hilbert space with inner product $\langle \cdot, \cdot \rangle$. Then the Banach algebra $B(H)$ of all bounded linear operators from H to itself is a C*-algebra in which T^* is the adjoint of T:

$$\langle Tx, y \rangle = \langle x, T^*y \rangle \qquad (x, y \in H)$$

and the identity $\|T^*T\| = \|T\|^2$ is well known. If $\dim H = n$, then $B(H)$ is just the algebra M_n of $n \times n$ complex matrices. The identity $\mathbf{1}$ in $B(H)$ is the identity operator on H.

Evidently, a closed *-*subalgebra* $\mathcal{A}$ of $B(H)$, that is, a subalgebra $\mathcal{A}$ of $B(H)$, closed in the norm topology of $B(H)$ and satisfying $x \in A \Rightarrow x^* \in A$, is a C*-algebra. Such an algebra is regarded as a '*concrete*' C*-algebra. In fact, every C*-algebra is (isomorphic to) a closed *-subalgebra of some $B(H)$. In other words, every '*abstract*' C*-algebra has a representation as a "*concrete*" C*-algebra.

Theorem 1.9 (Gelfand–Naimark Theorem). *Let $\mathcal{A}$ be a C*-algebra. Then there exists an isometric *-monomorphism $\pi : \mathcal{A} \longrightarrow B(H)$, for some Hilbert space H.*

Proof. Call a linear functional $f : \mathcal{A} \longrightarrow \mathbb{C}$ *positive*, in symbols $f \geq 0$, if $f(a^*a) \geq 0$ for all $a \in \mathcal{A}$. Let

$$N_f = \{a \in \mathcal{A} : f(a^*a) = 0\}.$$

The quotient $\mathcal{A}/N_f$ is an inner product space with the inner product

$$\langle a + N_f, b + N_f \rangle := f(b^*a).$$

Let H_f be the completion of $\mathcal{A}/N_f$. Then one can define a linear map

$$\pi_f : \mathcal{A} \longrightarrow B(H_f)$$

which satisfies

$$\pi_f(a)(b + N_f) = ab + N_f \qquad (b + N_f \in \mathcal{A}/N_f \subset H_f).$$

Take the direct sum of all the maps π_f induced by $f \geq 0$:

$$\pi = \bigoplus_{f \geq 0} \pi_f : \mathcal{A} \longrightarrow B\left(\bigoplus_{f \geq 0} H_f\right),$$

where $\bigoplus_{f \geq 0} H_f$ is the usual direct sum of the Hilbert spaces H_f:

$$\bigoplus_{f \geq 0} H_f = \left\{ (x_f)_{f \geq 0} : x_f \in H_f \text{ and } \sum_{f \geq 0} \|x_f\|^2 < \infty \right\}$$

and π is defined by

$$\left(\bigoplus_{f \geq 0} \pi_f\right)(a)((x_f)_{f \geq 0}) = (\pi_f(a)(x_f))_{f \geq 0} \qquad (a \in \mathcal{A}).$$

The map π is called the universal representation of $\mathcal{A}$ and is the required isometric *-monomorphism. □

The construction of the map π_f above is often called the *GNS-construction*, named after Gelfand, Naimark and Segal. More details of the construction will be given later.

Exercise 1.10. Given that $\mathcal{A}$ has an identity in *Theorem* 1.5, show that the spectrum $\Omega_{\mathcal{A}}$ of $\mathcal{A}$ is compact in the weak*-topology.

2. Spectral Theory

Let $\mathcal{A}$ be a C*-algebra. We define the *unit extension* of $\mathcal{A}$ to be the vector space direct sum $\mathcal{A}_1 = \mathcal{A} \oplus \mathbb{C}$, equipped with the following product and involution:

$$(a \oplus \lambda)(b \oplus \mu) = (ab + \lambda b + \mu a) \oplus (\lambda \mu), \quad (a \oplus \lambda)^* = a^* \oplus \overline{\lambda}.$$

Then $\mathcal{A}_1$ is a C*-algebra in the following norm

$$\|x\| = \sup\{\|xa\| : a \in \mathcal{A}, \|a\| \leq 1\} \qquad (x \in \mathcal{A}_1).$$

$\mathcal{A}_1$ is unital with identity $0 \oplus 1$.

Given a unital Banach algebra $\mathcal{A}$ and $a \in \mathcal{A}$, we define the *spectrum* of a to be the following subset of $\mathbb{C}$:

$$\sigma(a) = \{\lambda \in \mathbb{C} : \lambda\mathbf{1} - a \text{ is not invertible in } \mathcal{A}\}.$$

We will write λ for $\lambda\mathbf{1}$ if there is no confusion. The complement $\mathbb{C}\backslash\sigma(a)$ is called the *resolvent set* of a. If a C*-algebra $\mathcal{A}$ is not unital, then we define the *quasi-spectrum* of an element $a \in \mathcal{A}$ to be the following set:

$$\sigma'(a) = \{\lambda \in \mathbb{C} : \lambda\mathbf{1} - a \text{ is not invertible in } \mathcal{A}_1\}.$$

We always have $0 \in \sigma'(a)$! If $\mathcal{A}$ is unital, we have $\sigma'(a) = \sigma(a) \cup \{0\}$.

Proposition 2.1. *Let $\mathcal{A}$ be a Banach algebra and let $\omega : \mathcal{A} \longrightarrow \mathbb{C}$ be an algebra homomorphism. Then ω is continuous and $\|\omega\| \leq 1$.*

Proof.   There is nothing to prove if $\omega = 0$. Otherwise, extend ω to an algebra homomorphism $\widetilde{\omega}$ on the unit extension $\mathcal{A}_1$ of $\mathcal{A}$ by defining

$$\widetilde{\omega}(a + \alpha) = \omega(a) + \alpha \qquad (a + \alpha \in \mathcal{A}_1).$$

Then $\omega(x) - x$ is not invertible for each $x \in \mathcal{A}$ since $\omega(x) - x$ is in the kernel of $\widetilde{\omega}$. Hence we have $|\omega(x)| \leq \|x\|$ by Exercise 2.15. $\qquad\square$

Lemma 2.2. *Let $\mathcal{A}$ be a unital Banach algebra. The set $G(\mathcal{A})$ of invertible elements in $\mathcal{A}$ is open.*

Proof.   Let $a \in G(\mathcal{A})$. Then the open ball

$$B\left(a, \frac{1}{\|a^{-1}\|}\right) = \left\{x \in \mathcal{A} : \|x - a\| < \frac{1}{\|a^{-1}\|}\right\},$$

centred at a with radius $\frac{1}{\|a^{-1}\|}$, is contained in $G(\mathcal{A})$ since $x \in B(a, \frac{1}{\|a^{-1}\|})$ implies $\|\mathbf{1} - a^{-1}x\| = \|a^{-1}(a - x)\| < 1$ and hence $a^{-1}x$ is invertible by Exercise 2.15, which implies invertibility of x. $\qquad\square$

Proposition 2.3. *Let $\mathcal{A}$ be a unital Banach algebra and let $a \in \mathcal{A}$. Then the spectrum $\sigma(a)$ is a compact set in $\mathbb{C}$.*

Proof. The function $f : \lambda \in \mathbb{C} \mapsto (\lambda - a) \in \mathcal{A}$ is clearly continuous. Hence $\mathbb{C} \backslash \sigma(a) = f^{-1}(G(\mathcal{A}))$ is open.

If $|\lambda| > \|a\|$, then $\|\frac{a}{\lambda}\| < 1$ implies $1 - \frac{a}{\lambda}$, and hence $\lambda - a$, is invertible. Therefore $\sigma(a) \subset \{z \in \mathbb{C} : |z| \leq \|a\|\}$, that is, $\sigma(a)$ is bounded and compactness follows from the Heine–Borel theorem. $\qquad \square$

Proposition 2.4. *Let $\mathcal{A}$ be a unital Banach algebra and let $a \in \mathcal{A}$. Then $\sigma(a) \neq \emptyset$.*

Proof. Define the resolvent map $R : \mathbb{C} \backslash \sigma(a) \longrightarrow \mathcal{A}$ by

$$R(\lambda) = (\lambda - a)^{-1} \qquad (\lambda \in \mathbb{C}).$$

Then we have

$$\frac{R(\lambda) - R(\mu)}{\lambda - \mu} = -R(\mu)R(\lambda) \qquad (\lambda, \mu \in \mathbb{C})$$

and R is an $\mathcal{A}$-valued analytic function.

For each $\varphi \in \mathcal{A}^*$, the function $\varphi \circ R : \mathbb{C} \backslash \sigma(a) \longrightarrow \mathbb{C}$ is analytic. If $\sigma(a) = \emptyset$, then $\varphi \circ R$ is an entire function. Since

$$|\varphi \circ R(\lambda)| \leq \|\varphi\| \|R(\lambda)\| = \|\varphi\| |\lambda|^{-1} \|(1 - a/\lambda)^{-1}\| \longrightarrow 0$$

as $|\lambda| \to \infty$, we must have $\varphi \circ R$ identically 0, by Liouville theorem.

Since φ was arbitrary, we have $R = 0$ which is impossible. Hence $\sigma(a) \neq \emptyset$. $\qquad \square$

An algebra with identity is called a *division algebra* if every non-zero element in it is invertible.

Theorem 2.5 (Gelfand–Mazur Theorem). *Let $\mathcal{A}$ be a unital Banach algebra. If $\mathcal{A}$ is a division algebra, then $\mathcal{A} = \mathbb{C}1$.*

Definition 2.6. Let $\mathcal{A}$ be a C*-algebra and let $a \in \mathcal{A}$. The supremum

$$r(a) = \sup\{|\lambda| : \lambda \in \sigma'(a)\}$$

is called the *spectral radius* of a.

By the proof of Proposition 2.3, we have $r(a) \leq \|a\|$. In fact, we have the following useful formula for the spectral radius.

Theorem 2.7. *Let $a \in \mathcal{A}$. Then we have $r(a) = \lim_{n \to \infty} \|a^n\|^{1/n}$.*

Proof. We may assume that $\mathcal{A}$ has an identity. We first note that $\lambda \in \sigma(a)$ implies $\lambda^n \in \sigma(a^n)$ since

$$\lambda^n - a^n = (\lambda - a) \sum_{k=0}^{n-1} \lambda^k a^{n-1-k}.$$

It follows that $|\lambda|^n \leq \|a^n\|$ for all n and $r(a) \leq \lim_{n \to \infty} \inf \|a^n\|^{1/n}$.

For each $\varphi \in \mathcal{A}^*$, the complex function $\varphi \circ R$ is analytic in the domain

$$D = \{\lambda \in \mathbb{C} : |\lambda| > r(a)\} \subset \mathbb{C} \backslash \sigma(a).$$

By analyticity of $R(\lambda)$, it has the Laurent series

$$R(\lambda) = (\lambda \mathbf{1} - a)^{-1} = \sum_{n=0}^{\infty} \frac{a^n}{\lambda^{n+1}}$$

for $\lambda \in D$ and hence $\varphi \circ R(\lambda) = \sum_{n=0}^{\infty} \frac{\varphi(a^n)}{\lambda^{n+1}}$ which is the Laurent series of $\varphi \circ R$ in D. By convergence of the series, we have

$$\left| \frac{\varphi(a^n)}{\lambda^{n+1}} \right| \leq C_\varphi \qquad \text{for all } n,$$

for some constant C_φ depending on φ.

By the Uniform Boundedness Principle [15, Theorem 2.6], we must have

$$\frac{\|a^n\|}{|\lambda|^{n+1}} \leq C \qquad (\lambda \in D)$$

for all n. Hence $\|a^n\|^{1/n} \leq C^{1/n} |\lambda|^{1+\frac{1}{n}}$ for $|\lambda| > r(a)$, which yields $\lim_{n \to \infty} \sup \|a^n\|^{1/n} \leq r(a)$ and the proof is complete. $\qquad \square$

Definition 2.8. Let a be an element in a C*-algebra $\mathcal{A}$. It is called *self-adjoint* or *Hermitian* if $a^* = a$. It is called *normal* if $a^*a = aa^*$. It is called a *projection* if $a = a^* = a^2$. It is called *unitary* if $a^*a = aa^* = 1$, given that $\mathcal{A}$ is unital.

Every element $a \in \mathcal{A}$ can be written in the form

$$a = a_1 + ia_2,$$

where a_1 and a_2 are self-adjoint. In fact,

$$a_1 = \frac{1}{2}(a + a^*) \quad \text{and} \quad a_2 = \frac{1}{2i}(a - a^*).$$

Corollary 2.9. *Let a be a normal element in a C*-algebra $\mathcal{A}$. Then we have $\|a\| = r(a)$.*

Proof. We first show this for a self-adjoint element a. We have $\|a^2\| = \|a^*a\| = \|a\|^2$ and hence, by iteration,

$$r(a) = \lim_{n \to \infty} \|a^{2^n}\|^{1/2^n} = \|a\|.$$

For a normal element a, we have

$$r(a)^2 \leq \|a\|^2 = \|a^*a\| = \lim_{n \to \infty} \|(a^*a)^n\|^{1/n}$$

$$\leq \lim_{n \to \infty} \|(a^*)^n\|^{1/n} \|(a)^n\|^{1/n} = r(a)^2. \qquad \square$$

Proposition 2.10. *Let $\mathcal{A}$ be a unital C*-algebra. If $u \in \mathcal{A}$ is unitary, then $\sigma(u) \subset \{\lambda \in \mathbb{C} : |\lambda| = 1\}$. If $a \in \mathcal{A}$ is self-adjoint, then $\sigma(a) \subset \mathbb{R}$.*

Proof. We have $\|u\|^2 = \|u^*u\| = \|1\| = 1$ and since $u^* = u^{-1}$, the set

$$\sigma(u^*) = \{\overline{\lambda} : \lambda \in \sigma(u)\} = \sigma(u^{-1}) = \{\lambda^{-1} : \lambda \in \sigma(u)\}$$

is contained in the unit *disc* in $\mathbb{C}$, and hence $\sigma(u)$ is contained in the unit *circle* in $\mathbb{C}$.

Given a self-adjoint element $a \in \mathcal{A}$, the element

$$u = \exp(ia) = \sum_{n=0}^{\infty} \frac{(ia)^n}{n!}$$

is unitary since $u^* = \exp(-ia)$. We have

$$\{\exp i\lambda : \lambda \in \sigma(a)\} = \sigma(u) \subset \{\lambda \in \mathbb{C} : |\lambda| = 1\}$$

which implies $\sigma(a) \subset \mathbb{R}$. $\qquad \square$

Let $\mathcal{A}$ be a C*-algebra. A subspace $I \subset \mathcal{A}$ is called a *left ideal* if

$$a \in \mathcal{A} \quad \text{and} \quad x \in I \Longrightarrow ax \in I.$$

A *right ideal* of $\mathcal{A}$ is a subspace I satisfying

$$a \in \mathcal{A} \quad \text{and} \quad x \in I \Longrightarrow xa \in I.$$

We note that a left or right ideal is also a subalgebra of $\mathcal{A}$. We call I *proper* if $I \neq \mathcal{A}$. A *two-sided ideal* of $\mathcal{A}$ is a subspace which is both a left and a right ideal. If $\mathcal{A}$ is abelian, we speak simply of ideals rather than left or right ideals, in which case a *maximal ideal* is a proper ideal not properly contained in any proper ideal.

If I is a proper ideal in a unital $\mathcal{A}$, then I cannot contain any invertible element, that is $I \subset \mathcal{A} \backslash G(\mathcal{A})$. Since $G(\mathcal{A})$ is open by Lemma 2.2, the closure $\overline{I}$ is also contained in $\mathcal{A} \backslash G(\mathcal{A})$ and is therefore a proper ideal in $\mathcal{A}$. It follows that maximal ideals in $\mathcal{A}$ are closed.

There is a one-to-one correspondence between maximal ideals in an abelian C*-algebra $\mathcal{A}$ and the characters of $\mathcal{A}$.

Proposition 2.11. *Let $\mathcal{A}$ be a unital abelian C*-algebra with spectrum $\Omega_{\mathcal{A}}$. Then the mapping*

$$\omega \in \Omega_{\mathcal{A}} \mapsto \omega^{-1}(0) \subset \mathcal{A}$$

is a bijection onto the set of all maximal ideals of $\mathcal{A}$.

Proof. Let $\omega \in \Omega_{\mathcal{A}}$. Then $\omega^{-1}(0)$ is a proper ideal in $\mathcal{A}$ since $1 \notin \omega^{-1}(0)$. It is maximal because it has codimension 1.

Given a maximal ideal $M \subset \mathcal{A}$, the quotient Banach space $\mathcal{A}/M$ is a unital commutative Banach algebra in the product

$$(a + M)(b + M) := ab + M$$

and the quotient map

$$q : \mathcal{A} \longrightarrow \mathcal{A}/M$$

is an algebra homomorphism. By maximality of M, the Banach algebra $\mathcal{A}/M$ has no non-trivial ideal. Hence every non-zero element $[a] := a + M$ in $\mathcal{A}/M$ is invertible for otherwise, $[a]\,(\mathcal{A}/M)$ would be a non-trivial ideal. It follows from Mazur's theorem that there is an isomorphism $\varphi : \mathcal{A}/M \longrightarrow \mathbb{C}$ and hence $M = (\varphi \circ q)^{-1}(0)$ with $\varphi \circ q \in \Omega_{\mathcal{A}}$. $\square$

We can now give complete details of the proof of Theorem 1.5.

Proposition 2.12. *Let $\mathcal{A}$ be an abelian C*-algebra. Then the Gelfand map $\widehat{\ } : \mathcal{A} \longrightarrow C_0(\Omega_{\mathcal{A}})$ defined in Theorem 1.5 is an isometry and a $*$-map.*

Proof. If $\mathcal{A}$ is unital, then each non-invertible element $b \in \mathcal{A}$ is contained in some maximal ideal since $b\mathcal{A}$ is a proper ideal containing b and Zorn's lemma applies. It follows that

$$\alpha \in \sigma(a) \Leftrightarrow \alpha - a \in \omega^{-1}(0) \quad \text{for some } \omega \in \Omega_{\mathcal{A}}$$
$$\Leftrightarrow \alpha = \omega(a) = \widehat{a}(\omega) \quad \text{for some } \omega \in \Omega_{\mathcal{A}}$$

and we have

$$\sigma(a) = \widehat{a}(\Omega_{\mathcal{A}}). \tag{2.1}$$

If $\mathcal{A}$ is non-unital, then the quasi-spectrum $\sigma'(a)$ is the spectrum $\sigma_{\mathcal{A}_1}(a)$ of a in the unit extension $\mathcal{A}_1$ of $\mathcal{A}$ and we have

$$\sigma'(a) = \sigma_{\mathcal{A}_1}(a) = \widehat{a}(\Omega_{\mathcal{A}_1}) = \widehat{a}(\Omega_{\mathcal{A}}), \tag{2.2}$$

where $\Omega_{\mathcal{A}_1} = \Omega_{\mathcal{A}} \cup \{\omega_0\}$ and $\omega_0(a \oplus \beta) = \beta$ for $a \oplus \beta \in \mathcal{A}_1$.

In both cases, we have $\|a\| = r(a) = \|\widehat{a}\|$ since a is normal in the abelian algebra $\mathcal{A}$.

To see that the Gelfand map is a $*$-map, let $a \in \mathcal{A}$ and $a = a_1 + ia_2$ where a_1 and a_2 are self-adjoint. For each $\omega \in \Omega$, we have

$$\widehat{a^*}(\omega) = \omega(a^*) = \omega(a_1 - ia_2) = \omega(a_1) - i\omega(a_2),$$
$$= \overline{\omega(a_1) + i\omega(a_2)} = \overline{\omega(a)} = \overline{\widehat{a}}(\omega),$$

where $\omega(a_i) = \widehat{a_i}(\omega) \in \sigma(a_i) \subset \mathbb{R}$ for $i = 1, 2$. This proves $\widehat{a^*} = \overline{\widehat{a}}$. □

Remark 2.13. The above proposition shows that

$$\|a\| = \sup\{|\omega(a)| : \omega \in \Omega_{\mathcal{A}}\}$$

for each element a in an abelian C*-algebra $\mathcal{A}$.

To conclude the proof of Theorem 1.5, we observe that the Gelfand map $\widehat{} : \mathcal{A} \longrightarrow C(\Omega_{\mathcal{A}})$ is surjective if $\mathcal{A}$ is unital since $\widehat{\mathcal{A}}$ is a closed $*$-subalgebra of $C(\Omega_{\mathcal{A}})$ containing constant functions on $\Omega_{\mathcal{A}}$ and separating points of $\Omega_{\mathcal{A}}$ which imply $\widehat{\mathcal{A}} = C(\Omega)$ by the Stone–Weierstrass theorem.

If $\mathcal{A}$ is non-unital, the Gelfand map $\widehat{} : \mathcal{A} \longrightarrow C_0(\Omega_{\mathcal{A}})$ is also surjective. Indeed, each $f \in C_0(\Omega_{\mathcal{A}})$ can be extended to a continuous function $\overline{f}$ on

$$\Omega_{\mathcal{A}_1} = \Omega_{\mathcal{A}} \cup \{\omega_0\}$$

by defining $\overline{f}(\omega_0) = 0$, and therefore, the surjectivity of the Gelfand map on $\mathcal{A}_1$ implies that $\overline{f} = \widehat{b}$ for some $b = a \oplus \beta \in \mathcal{A}_1$; but $\beta = \omega_0(a \oplus \beta) = \omega_0(b) = \overline{f}(\omega_0) = 0$. Hence $f = \widehat{a} \in \widehat{\mathcal{A}}$.

Exercise 2.14. Verify that the unit extension $\mathcal{A}_1$ is a C*-algebra.

Exercise 2.15. Let $\mathcal{A}$ be a unital Banach algebra and let $a \in \mathcal{A}$ satisfy $\|a\| < 1$. Show that $1 - a$ is invertible in $\mathcal{A}$.

Exercise 2.16. Prove the Gelfand–Mazur theorem.

3. Functional Calculus

Given a non-empty subset S of a C*-algebra $\mathcal{A}$, the smallest C*-subalgebra of $\mathcal{A}$ containing S clearly exists, by taking the intersection of all C*-subalgebras containing S. We call it the C*-*algebra generated by S* in $\mathcal{A}$. We begin with the following two fundamental results which enable us to derive properties of C*-algebras by reduction to the commutative case.

Theorem 3.1. *Let $\mathcal{A}$ be a unital C^*-algebra and let $a \in \mathcal{A}$ be normal. Then the C^*-subalgebra $C(a, 1)$ generated by a and 1 is isometrically *-isomorphic to the abelian C^*-algebra $C(\sigma(a))$ of complex continuous functions on the spectrum $\sigma(a)$ of a in $\mathcal{A}$.*

Proof.   Since a is normal, the C*-algebra $C(a, 1)$, consisting of polynomials in a and a^*, as well as their limits, is abelian and can be identified with the C*-algebra $C(\Omega)$ of continuous functions on the spectrum Ω of $C(a, 1)$, via the Gelfand transform $\widehat{}$. Let $\sigma_C(a)$ be the spectrum of a in $C(a, 1)$. By (2.1) in the proof of Proposition 2.12, we have $\sigma_C(a) = \widehat{a}(\Omega)$ and the map

$$\omega \in \Omega \mapsto \widehat{a}(\omega) \in \sigma_C(a)$$

is a homeomorphism since $C(a, 1)$ is generated by a. It follows that

$$f \in C(\sigma_C(a)) \mapsto f \circ \widehat{a} \in C(\Omega)$$

is an isometric *-isomorphism.

It remains to show that $\sigma(a) = \sigma_C(a)$. Evidently, we have $\sigma(a) \subset \sigma_C(a)$. Let $\alpha \in \sigma_C(a)$. Regard $f = a - \alpha 1$ as a function in $C(\sigma_C(a))$. For any $\varepsilon > 0$, the set $K = \{x \in \sigma_C(a) : |f(x)| \geq \varepsilon\}$ is compact in $\sigma_C(a)$ and we can find a function $g \in C(\sigma_C(a))$ satisfying $0 \leq g \leq 1$, $g(\alpha) = 1$ and $g(K) = \{0\}$, by Urysohn Lemma, so that $\|(a - \alpha)g\| \leq \varepsilon$ and hence $\alpha \in \sigma(a)$, for if $b(a - \alpha) = 1$ for some $b \in \mathcal{A}$, we can choose $g \in C(\sigma_C(a))$ with $\|g\| = 1$ and $\|(a - \alpha)g\| < \|b\|^{-1}$, giving a contradiction. This proves $\sigma(a) = \sigma_C(a)$. $\square$

Theorem 3.2. *Let $\mathcal{A}$ be a C^*-algebra and let $a \in \mathcal{A}$ be normal. Then the C^*-subalgebra $C(a)$ generated by a is isometrically *-isomorphic to the abelian C^*-algebra $C_0(\sigma'(a)\backslash\{0\})$ of complex continuous functions on $\sigma'(a)\backslash\{0\}$, vanishing at infinity, where $\sigma'(a)$ is the quasi-spectrum of a in $\mathcal{A}$.*

Proof. Normality of a implies that $C(a)$ is an abelian C*-algebra and we have $C(a) \simeq C_0(\Omega)$ via the Gelfand transform, where Ω is the spectrum of $C(a)$. Since a generates $C(a)$, we have $\omega(a) \neq 0$ for all $\omega \in \Omega$ and by (2.2), the map

$$\omega \in \Omega \mapsto \widehat{a}(\omega) \in \sigma'_{C(a)}(a)\backslash\{0\}$$

is a homeomorphism and we have $C(a) \simeq C_0(\sigma'_{C(a)}(a)\backslash\{0\})$. As in the proof of the above theorem, the non-zero quasi-spectrum $\sigma'(a)\backslash\{0\}$ coincides with the non-zero quasi-spectrum $\sigma'_{C(a)}(a)\backslash\{0\}$ of a in $C(a)$. Hence we have $C(a) \simeq C_0(\sigma'(a)\backslash\{0\})$. $\qquad\square$

Let $a \in \mathcal{A}$ be a self-adjoint element and let $C(a) \subset \mathcal{A}$ be the C*-algebra generated by a. Denote by $\varphi : C_0(\sigma'(a)\backslash\{0\}) \longrightarrow C(a)$ the *-isomorphism in Theorem 3.2. For each $f \in C_0(\sigma'(a)\backslash\{0\})$, we write $f(a) = \varphi(f)$. Evidently $a = \varphi(\iota) = \iota(a)$ where $\iota \in C_0(\sigma'(a)\backslash\{0\})$ is the identity map $\iota : \sigma'(a)\backslash\{0\} \longrightarrow \sigma'(a)\backslash\{0\}$. Since $\sigma'(a) \subset \mathbb{R}$, we can define $f_1, f_2, f_3, f_4 \in C_0(\sigma'(a)\backslash\{0\})$ to be the following real-valued functions

$$f_1(\lambda) = \lambda \vee 0, \quad f_2(\lambda) = \lambda \wedge 0$$
$$f_3(\lambda) = |\lambda|, \quad f_4(\lambda) = \sqrt{\lambda} \quad \text{if } \sigma'(a) \subset [0, \infty).$$

Write $a_+ = f_1(a)$, $a_- = f_2(a)$, $|a| = f_3(a)$ and $a^{1/2} = f_4(a)$. Then we have $a = a_+ - a_-$, $|a| = a_+ + a_-$ and $a_+ a_- = 0$. Such calculation is an example of what is often called *functional calculus*.

Lemma 3.3. *If $\varphi \in \mathcal{A}^*$ satisfies $\varphi(a^*a) = 0$ for all $a \in \mathcal{A}$, then $\varphi = 0$.*

Proof. By a remark after Definition 2.8, it suffices to show that $\varphi(a) = 0$ for each self-adjoint element $a \in \mathcal{A}$. By functional calculus, we have $\varphi(a) = \varphi(a_+ - a_-) = \varphi(a_+^{1/2}a_+^{1/2}) - \varphi(a_-^{1/2}a_-^{1/2}) = 0$. $\qquad\square$

A self-adjoint element $a \in \mathcal{A}$ is called *positive*, written $a \geq 0$ or $0 \leq a$ if $\sigma'(a) \subset [0, \infty)$. We see from the above that $a \geq 0$ if and only if $a = a^+$. Moreover, it can be shown that $a \geq 0$ if and only if $a = b^*b$ for some $b \in \mathcal{A}$ (cf. [17, p. 23]). The self-adjoint elements in $\mathcal{A}$ form a real closed subspace $\mathcal{A}_{sa}$ of $\mathcal{A}$. The positive elements of $\mathcal{A}$ form a cone in $\mathcal{A}_{sa}$ and induces a partial ordering $\leq$, where $x \leq y$ in $\mathcal{A}_{sa}$ if $x - y \leq 0$. We also write $y \geq x$ for this.

Given a self-adjoint element a in a unital C*-algebra $\mathcal{A}$, by considering a as the identity function ι on the spectrum $\sigma(a) \subset [0, \infty)$ and $\mathbf{1}$ as the constant function on $\sigma(a)$ with value 1, we see that

$$-\|a\|\mathbf{1} \leq a \leq \|a\|\mathbf{1}$$

and also, $-t\mathbf{1} \leq a \leq t\mathbf{1}$ implies $\|a\| \leq t$ for any $t \geq 0$. Given $0 \leq x \leq y$ in $\mathcal{A}$, we have $0 \leq y \leq \|y\|\mathbf{1}$ and therefore $0 \leq x \leq \|y\|\mathbf{1}$. It follows that $\|x\| \leq \|y\|$.

It is often useful to make use of the identity in a C*-algebra if it has one. If there is no identity, one can add one by forming the unit extension of the C*-algebra in question, as we have done so in some previous arguments. Another useful device is the *approximate identity* which exists in every C*-algebra. A net (u_α) of positive elements in a C*-algebra $\mathcal{A}$ is called an *approximate identity* if it is increasing and $\|u_\alpha\| \leq 1$ for all α such that

$$x = \lim_\alpha x u_\alpha = \lim_\alpha u_\alpha x \qquad (x \in \mathcal{A}).$$

Theorem 3.4. *Every C*-algebra contains an approximate identity.*

Proof. See, for example, [14, p. 11]. □

4.   Homomorphisms of C*-algebras

A simple spectral analysis shows that a $*$-homomorphism $\varphi : \mathcal{A} \longrightarrow \mathcal{B}$ is automatically continuous. In fact, it is contractive.

Theorem 4.1. *Let $\mathcal{A}$ and $\mathcal{B}$ be C*-algebras and let $\varphi : \mathcal{A} \longrightarrow \mathcal{B}$ be a $*$-homomorphism. Then we have $\|\varphi(a)\| \leq \|a\|$ for each $a \in \mathcal{A}$.*

Proof. Since φ is a homomorphism, we have

$$\sigma'(a) \supset \sigma'(\varphi(a))$$

and hence the following inequalities for the spectral radii:

$$r(\varphi(a)) \leq r(a) \leq \|a\|$$

for all $a \in \mathcal{A}$. Noting that the spectral radius of a self-adjoint element coincides with its norm, we obtain, for each $a \in \mathcal{A}$,

$$\|\varphi(a)\|^2 = \|\varphi(a)\varphi(a)^*\| = \|\varphi(aa^*)\| = r(\varphi(aa^*)) \le \|aa^*\| = \|a\|^2,$$

which completes the proof. □

Corollary 4.2. *Let $\varphi : \mathcal{A} \longrightarrow \mathcal{B}$ be a *-isomorphism from $\mathcal{A}$ onto a C*-algebra $\mathcal{B}$. Then φ is an isometry, that is, $\|\varphi(a)\| = \|a\|$ for all $a \in \mathcal{A}$.*

Proof. Apply the above theorem to the *-isomorphism φ and its inverse φ^{-1}. □

One can show further that a *-isomorphism from a C*-algebra $\mathcal{A}$ *into* a C*-algebra $\mathcal{B}$ is an isometry. We recall that the *dual linear map φ^** : $F^* \longrightarrow E^*$ of a continuous linear map $\varphi : E \longrightarrow F$ between Banach spaces is defined by

$$\varphi^*(\omega)(x) = \omega(\varphi(x)) \qquad (\omega \in F^*, x \in E).$$

Theorem 4.3. *Let $\varphi : \mathcal{A} \longrightarrow \mathcal{B}$ be a *-isomorphism from a C*-algebra $\mathcal{A}$ into a C*-algebra $\mathcal{B}$. Then φ is an isometry.*

Proof. We do not know, *a priori*, that the image $\varphi(\mathcal{A})$ is a C*-algebra since in general, a continuous image of a Banach space need not be a Banach space. Therefore the above arguments cannot be applied directly to the inverse of φ.

Let $a \in \mathcal{A}$. By considering the C*-subalgebras generated by a and $\varphi(a)$, we may assume both $\mathcal{A}$ and $\mathcal{B}$ are abelian. By adding the identity, we may assume further that $\mathcal{A}$ and $\mathcal{B}$ are unital, and that $\varphi(\mathbf{1}) = \mathbf{1}$.

Identify $\mathcal{A}$ with $C(\Omega_{\mathcal{A}})$ and $\mathcal{B}$ with $C(\Omega_{\mathcal{B}})$ via the Gelfand transform. As usual, we can identify $\Omega_{\mathcal{A}}$ as a weak* compact subspace of the dual $C(\Omega_{\mathcal{A}})^*$ by the evaluation map:

$$\omega \in \Omega_{\mathcal{A}} \mapsto \varepsilon_\omega \in C(\Omega_{\mathcal{A}})^*, \quad \text{where } \varepsilon_\omega(a) = a(\omega) \quad (a \in C(\Omega_{\mathcal{A}}), \omega \in \Omega_{\mathcal{A}}).$$

Likewise $\Omega_{\mathcal{B}} \subset C(\Omega_{\mathcal{B}})^*$.

Since $\Omega_{\mathcal{A}}$ and $\Omega_{\mathcal{B}}$ are spectra of $\mathcal{A}$ and $\mathcal{B}$ respectively, and since φ is an algebra isomorphism, the dual map $\varphi^* : C(\Omega_{\mathcal{B}})^* \longrightarrow C(\Omega_{\mathcal{A}})^*$ carries $\Omega_{\mathcal{B}}$ into $\Omega_{\mathcal{A}}$, and $\varphi^*(\Omega_{\mathcal{B}})$ is weak* compact, by continuity of φ^*, and hence weak* closed.

We claim that $\varphi^*(\Omega_\mathcal{B}) = \Omega_\mathcal{A}$. Indeed, if $\varphi^*(\Omega_\mathcal{B}) \neq \Omega_\mathcal{A}$, then we can find two non-zero functions $f, g \in C(\Omega_\mathcal{A})$ such that $fg = 0$ and $g(\chi) = 1$ for all $\chi \in \varphi^*(\Omega_\mathcal{B})$. By the Gelfand representation, we can then find non-zero $a, b \in \mathcal{A}$ such that $ab = 0$ and $\chi(b) = 1$ for all $\chi \in \varphi^*(\Omega_\mathcal{B})$. Since $\varphi(a) \neq 0$, there exists $\omega \in \Omega_\mathcal{B}$, such that $\omega(\varphi(a)) \neq 0$.

It follows from $\varphi(a)\varphi(b) = 0$ that

$$0 = \omega(\varphi(a)\varphi(b)) = \omega(\varphi(a))\omega(\varphi(b)) = \omega(\varphi(a))\varphi^*(\omega)(b) = \omega(\varphi(a)) \neq 0,$$

which is a contradiction. Therefore $\varphi^*(\Omega_\mathcal{B}) = \Omega_\mathcal{A}$ and

$$\|\varphi(a)\| = \sup\{|\varphi(a)(\omega)| : \omega \in \Omega_\mathcal{B}\} = \sup\{|\varphi^*(\omega)(a)| : \omega \in \Omega_\mathcal{B}\}$$
$$= \sup\{|\chi(a)| : \chi \in \Omega_\mathcal{A}\} = \|a\|. \qquad \square$$

Exercise 4.4 (Uniqueness of C*-norm). Let $(\mathcal{A}, \|\cdot\|)$ be a C*-algebra and let $\|\cdot\|_1$ be a norm on $\mathcal{A}$ such that $(\mathcal{A}, \|\cdot\|_1)$ is a C*-algebra. Show that $\|\cdot\| = \|\cdot\|_1$.

Exercise 4.5. Is the converse of *Corollary* 4.2 true?

Exercise 4.6. Is the converse of *Theorem* 4.3 true?

5.   States and Representations

Let $\mathcal{A}$ be a C*-algebra. A linear functional $\varphi : \mathcal{A} \longrightarrow \mathbb{C}$ is called *positive* if $\varphi(a^*a) \geq 0$ for all $a \in \mathcal{A}$.

Lemma 5.1. *A positive linear functional φ on a unital C*-algebra $\mathcal{A}$ satisfies*

 (i) $\varphi(x^*) = \overline{\varphi(x)}$ $(x \in \mathcal{A})$;
 (ii) $|\varphi(x^*y)|^2 \leq \varphi(x^*x)\varphi(y^*y)$ $(x, y \in \mathcal{A})$;
 (iii) φ *is continuous and* $\|\varphi\| = \varphi(\mathbf{1})$;
 (iv) $|\varphi(y^*xy)| \leq \|x\|\varphi(y^*y)$ $(x, y \in \mathcal{A})$;
 (v) $\|\varphi + \psi\| = \|\varphi\| + \|\psi\|$ *if ψ is also a positive linear functional on $\mathcal{A}$.*

Proof.   By positivity, φ induces a positive semidefinite sesquilinear form, i.e. a semi inner product, $\langle \cdot, \cdot \rangle : (x, y) \in \mathcal{A} \times \mathcal{A} \mapsto \varphi(y^*x)$. Hence (i) follows from Hermitian symmetry $\langle x, y \rangle = \overline{\langle y, x \rangle}$, and (ii) is the Schwarz inequality.

To see (iii), we first note that (ii) implies

$$|\varphi(x)|^2 \le \varphi(x^*x)\varphi(\mathbf{1}) \qquad (x \in \mathcal{A}).$$

The element $a = x^*x$ is self-adjoint and the C*-subalgebra $C(a, \mathbf{1})$ of $\mathcal{A}$, generated by a and $\mathbf{1}$, is identified with complex continuous functions on a compact Hausdorff space $\sigma(a)$, by Theorem 3.1.

If $\|x\| \le 1$, then $\|a\| = \|x\|^2 \le 1$ and we must have, as functions on $\sigma(a)$, that $-1 \le a \le 1$ and hence $\mathbf{1} - a = b^*b$ for some $b \in C(a, \mathbf{1})$. Positivity of φ gives $\varphi(\mathbf{1} - x^*x) = \varphi(b^*b) \ge 0$ and $\varphi(x^*x) \le \varphi(\mathbf{1})$. It follows that

$$\sup\{|\varphi(x)| : \|x\| \le 1\} \le \varphi(\mathbf{1})$$

and φ is continuous with $\|\varphi\| \le \varphi(\mathbf{1})$. Hence $\|\varphi\| = \varphi(\mathbf{1})$ since $\|\mathbf{1}\| = 1$. $\qquad\square$

Remark 5.2. The assertions (i) and (ii) above do not require the identity of $\mathcal{A}$. Also (i) implies $\varphi(x) \in \mathbb{R}$ if x is self-adjoint.

Definition 5.3. A positive linear functional φ on a C*-algebra $\mathcal{A}$ is called a *state* if $\|\varphi\| = 1$.

Lemma 5.4. *Let $\mathcal{A}$ be a unital C*-algebra and let $\mathcal{B}$ be a C*-subalgebra containing the identity $\mathbf{1}$ of $\mathcal{A}$. Then every state φ on $\mathcal{B}$ extends to a state on $\mathcal{A}$, that is, there is a state $\widetilde{\varphi}$ on $\mathcal{A}$ such that $\widetilde{\varphi}|_{\mathcal{B}} = \varphi$.*

Proof. By the Hahn–Banach theorem [15, Theorem 3.3], φ extends to a linear functional $\widetilde{\varphi}$ on $\mathcal{A}$ with $\|\widetilde{\varphi}\| = \|\varphi\| = 1$. We have $\widetilde{\varphi}(\mathbf{1}) = 1$. We need to show $\widetilde{\varphi}(x^*x) \ge 0$ for all $x \in \mathcal{A}$.

We first show that $\widetilde{\varphi}(a) \in \mathbb{R}$ if $a = a^*$. Indeed, if $\widetilde{\varphi}(a) = \alpha + i\beta$ with $\alpha, \beta \in \mathbb{R}$ and $\beta \ne 0$, then the element

$$y = \beta^{-1}(a - \alpha\mathbf{1})$$

is self-adjoint and $\widetilde{\varphi}(y) = i$ and hence for all $r \in \mathbb{R}$, we have

$$(r + 1)^2 = |i + ri|^2 = |\widetilde{\varphi}(y + ri)|^2 \le \|y + ri\|^2$$
$$= \|(y + ri)^*(y + ri)\| = \|y^2 + r^2\| \le \|y\|^2 + r^2$$

which is impossible.

To show $\widetilde{\varphi}(x^*x) \geq 0$, we may assume $\|x\| \leq 1$, by linearity of $\widetilde{\varphi}$. The C*-subalgebra $C(x^*x, \mathbf{1})$ generated by x^*x and $\mathbf{1}$ identifies with the algebra $C(\Omega)$ of complex continuous functions on a compact Hausdorff space Ω, and $\mathbf{1}$ with the constant function on Ω taking value 1. As a function in $C(\Omega)$, the element x^*x has supremum norm at most 1 and hence $\|\mathbf{1} - x^*x\| \leq 1$. It follows that $1 \geq \widetilde{\varphi}(\mathbf{1} - x^*x) = \widetilde{\varphi}(\mathbf{1}) - \widetilde{\varphi}(x^*x) = 1 - \widetilde{\varphi}(x^*x)$ and $\widetilde{\varphi}(x^*x) \geq 0$, where $\widetilde{\varphi}(\mathbf{1} - x^*x) \in \mathbb{R}$ because $\mathbf{1} - x^*x$ is self-adjoint. $\square$

Lemma 5.5. *Let $\mathcal{A}$ be a C*-algebra and $a \in \mathcal{A}$. If $\varphi(a^*a) = 0$ for every state φ of $\mathcal{A}$, then $a = 0$.*

Proof. First, assume $\mathcal{A}$ has an identity $\mathbf{1}$. Let

$$\,\widehat{}\, : C(a^*a, \mathbf{1}) \longrightarrow C(\Omega)$$

be the Gelfand transform identifying the C*-subalgebra generated by a^*a and $\mathbf{1}$ with the algebra $C(\Omega)$ of complex continuous functions on the spectrum Ω of a^*a, shown in Theorem 3.1.

Each $\omega \in \Omega$ induces a state $\psi = \varepsilon_\omega \circ \,\widehat{}\,$ on $C(a^*a, \mathbf{1})$, where

$$\varepsilon_\omega(f) = f(\omega) \quad (f \in C(\Omega)).$$

By Lemma 5.4, ψ extends to a state φ of $\mathcal{A}$ and hence $\psi(a^*a) = \varphi(a^*a) = 0$ which gives $\widehat{a^*a}(\omega) = 0$.

Since $\omega \in \Omega$ was arbitrary, we have $\widehat{a^*a} = 0$ and therefore $a^*a = 0$ as well as $a = 0$.

Now, if $\mathcal{A}$ lacks an identity, we consider its unit extension $\mathcal{A}_1$. Each state ψ of $\mathcal{A}_1$ restricts to a state φ of $\mathcal{A}$ and by hypothesis, we have $\psi(a^*a) = \varphi(a^*a) = 0$. It follows that $a = 0$ by the above conclusion for the unital case. $\square$

Definition 5.6. Let $\mathcal{A}$ be a C*-algebra and let H be a Hilbert space. A *representation* π of $\mathcal{A}$ on H is a *-algebra homomorphism $\pi : \mathcal{A} \longrightarrow B(H)$, in other words, π is a linear map from $\mathcal{A}$ into $B(H)$ satisfying

$$\pi(ab) = \pi(a)\pi(b) \quad \text{and} \quad \pi(a^*) = \pi(a)^*$$

for all $a, b \in \mathcal{A}$. A representation π is called *faithful* if it is injective.

Two representations $\pi : \mathcal{A} \longrightarrow B(H)$ and $\tau : \mathcal{A} \longrightarrow B(K)$ are said to be *(unitarily) equivalent*, in symbols: $\pi \simeq \tau$, if there is a surjective linear isometry $u : H \longrightarrow K$ such that

$$u\pi(a) = \tau(a)u \quad (a \in \mathcal{A}).$$

We call u an *intertwining operator* between π and τ.

Let $\pi : \mathcal{A} \longrightarrow B(H)$ be a representation of $\mathcal{A}$. Given a closed subspace $K \subset H$ *invariant* under $\pi(\mathcal{A})$, that is, $\pi(\mathcal{A})(K) \subset K$, we can define a representation $\pi_K : \mathcal{A} \longrightarrow B(K)$ by restriction:

$$\pi_K(a) = \pi(a)|_K \qquad (a \in \mathcal{A}).$$

The representation π_K is called a *sub-representation* of π.

Definition 5.7. A representation $\pi : \mathcal{A} \longrightarrow B(H)$ is called *irreducible* if $\pi(\mathcal{A})$ has no (closed) invariant subspace other than $\{0\}$ and H.

Let $\pi : \mathcal{A} \longrightarrow B(H)$ be a representation and let $p \in B(H)$ be a projection. Then the range space $p(H)$ of p is an invariant subspace of $\pi(\mathcal{A})$ if and only if p commutes with every element in $\pi(\mathcal{A})$. Indeed, if p commutes with $\pi(\mathcal{A})$, then

$$\pi(\mathcal{A})p(H) = p\pi(\mathcal{A})(H) \subset p(H).$$

Conversely, the invariance of $p(H)$ implies that $p\pi(a)p\eta = \pi(a)p\eta$ for every $\eta \in H$. On the other hand, we have

$$p\pi(a)\eta = p\pi(a)(p\eta + (1-p)\eta) = p\pi(a)p\eta + p\pi(a)(1-p)\eta = p\pi(a)p\eta$$

since $H = p(H) \oplus (1-p)(H)$ and $\pi(\mathcal{A})(1-p)H \subset (1-p)H$.

Given a representation $\pi : \mathcal{A} \longrightarrow B(H)$ and a vector $\xi \in H$, it is easily verified that the function $\varphi : \mathcal{A} \longrightarrow \mathbb{C}$ defined by

$$\varphi(a) = \langle \pi(a)\xi, \xi \rangle \qquad (a \in \mathcal{A})$$

is a positive linear functional. Conversely, every positive functional φ of $\mathcal{A}$ induces a representation π_φ of $\mathcal{A}$ by the Gelfand–Naimark–Segal construction described below.

Let $\varphi : \mathcal{A} \longrightarrow \mathbb{C}$ be a positive linear functional. Define

$$N_\varphi = \{a \in \mathcal{A} : \varphi(a^*a) = 0\}$$

which is called the *left kernel* of φ. The Schwarz inequality in Lemma 5.1(ii) implies that N_φ is a closed subspace of $\mathcal{A}$. One can also show by Lemma 5.1(iv) that N_φ is a left ideal of $\mathcal{A}$. This is not needed below.

The quotient $\mathcal{A}/N_\varphi$ is an inner product space with the inner product

$$\langle a + N_\varphi, b + N_\varphi \rangle := \varphi(b^*a)$$

since $\langle a + N_\varphi, a + N_\varphi \rangle = 0$ if and only if $a \in N_\varphi$. Denote the inner product norm in A/N_φ by

$$\|a + N_\varphi\|_\varphi = \varphi(a^*a)^{1/2}.$$

Let H_φ be the completion of A/N_φ. We can define a map $\pi_\varphi : A \longrightarrow B(H_\varphi)$ satisfying

$$\pi_\varphi(a)(b + N_\varphi) = ab + N_\varphi \qquad (a \in A, b + N_\varphi \in A/N_\varphi).$$

Indeed, for $a \in A$, the above formula defines a bounded linear operator $\pi_\varphi(a) : A/N_\varphi \longrightarrow A/N_\varphi$ since

$$\|\pi_\varphi(a)(b+N_\varphi)\|_\varphi^2 = \|ab+N_\varphi\|_\varphi^2 = \varphi(b^*a^*ab) \le \|a\|^2\varphi(b^*b) = \|a\|^2\|b+N_\varphi\|_\varphi^2.$$

Hence $\pi_\varphi(a)$ can be extended to a bounded linear operator on the completion H_φ, and is still denoted by $\pi_\varphi(a)$.

We see that π_φ is a $*$-homomorphism. Hence it is continuous and $\|\pi_\varphi\| \le 1$ by Theorem 4.1. If A has identity $\mathbf{1}$, then we have

$$\varphi(a) = \langle \pi_\varphi(a)\xi_\varphi, \xi_\varphi \rangle \qquad (a \in A), \tag{5.1}$$

where $\xi_\varphi = \mathbf{1} + N_\varphi$. In this case, $\pi_\varphi(A)\xi_\varphi = A/N_\varphi$ and is dense in H_φ. If A has no identity, one can still find a vector $\xi_\varphi \in H_\varphi$ satisfying (5.1) with $\pi_\varphi(A)\xi_\varphi$ dense in H_φ. We refer to [17, p. 39] for a proof.

We call the representation π_φ constructed above the GNS-*representation* of φ, and ξ_φ a *cyclic vector* for π_φ.

Definition 5.8. Let A be a C*-algebra. The set $S(A)$ of all states of A is called the *state space* of A.

If A is unital, Lemma 5.1 implies that its state space $S(A)$ is a weak* closed convex subset of the dual ball $\{\varphi \in A^* : \|\varphi\| \le 1\}$ and is therefore weak* compact.

We are now ready to show that the direct sum $\bigoplus_\varphi \pi_\varphi$ of all the representations π_φ induced by the states φ of A is a $*$-isomorphism from A into $B(\bigoplus_\varphi H_\varphi)$. We first show that the mapping

$$\pi = \bigoplus_{\varphi \in S(A)} \pi_\varphi : A \longrightarrow B\left(\bigoplus_{\varphi \in S(A)} H_\varphi\right)$$

defined by

$$\pi(a)(\oplus_\varphi x_\varphi) = \left(\bigoplus_{\varphi \in S(\mathcal{A})} \pi_\varphi \right)(a)(\oplus x_\varphi) = \bigoplus_{\varphi \in S(\mathcal{A})} \pi_\varphi(a)(x_\varphi) \qquad (a \in \mathcal{A})$$

is indeed well defined. This follows from

$$\left\| \bigoplus_{\varphi \in S(\mathcal{A})} \pi_\varphi(a)(x_\varphi) \right\|^2 = \sum_{\varphi \in S(\mathcal{A})} \|\pi_\varphi(a)(x_\varphi)\|^2 \leq \sum_{\varphi \in S(\mathcal{A})} \|\pi_\varphi(a)\|^2 \|(x_\varphi)\|^2$$

$$\leq \|a\|^2 \sum_{\varphi \in S(\mathcal{A})} \|(x_\varphi)\|^2 = \|a\|^2 \| \oplus x_\varphi \|^2$$

which also implies $\|\pi(a)\| \leq \|a\|$. In fact, π is an isometry since it is a *-isomorphism and Theorem 4.3 applies. Indeed, it is clear that π is a *-homomorphism. If $\pi(a) = 0$, then $\pi_\varphi(a) = 0$ for all $\varphi \in S(\mathcal{A})$. Hence

$$\varphi(aa^*aa^*) = \|aa^* + N_\varphi\|^2 = \|\pi_\varphi(a)(a^* + N_\varphi)\|^2 = 0$$

for all $\varphi \in S(\mathcal{A})$. By Lemma 5.5, we have $aa^* = 0$ and $a = 0$. This shows that π is a *-monomorphism and completes the proof of Theorem 1.9.

Definition 5.9. Given two linear functionals ψ and φ of a C*-algebra $\mathcal{A}$. We write $\psi \leq \varphi$ if $\psi(a^*a) \leq \varphi(a^*a)$ for all $a \in \mathcal{A}$.

Definition 5.10. A positive linear functional φ of $\mathcal{A}$ is called *pure* if for any positive linear functional ψ of $\mathcal{A}$ satisfying $\psi \leq \varphi$, we have $\psi = \alpha\varphi$ for some $\alpha \geq 0$. Note that $\alpha \leq 1$.

Theorem 5.11. *Let $\mathcal{A}$ be a C*-algebra and let φ be a state of $\mathcal{A}$. The following conditions are equivalent.*

(i) *φ is a pure state.*
(ii) *The GNS-representation π_φ is irreducible.*

Proof. (i) $\Longrightarrow$ (ii) Let φ be a pure state and let $\pi_\varphi : \mathcal{A} \longrightarrow B(H_\varphi)$ be the GNS-representation such that

$$\varphi(a) = \langle \pi_\varphi(a)\xi_\varphi, \xi_\varphi \rangle \qquad (a \in \mathcal{A})$$

for some cyclic vector $\xi_\varphi \in H_\varphi$ and $\pi_\varphi(\mathcal{A})\xi_\varphi$ is dense in H_φ.

Let $K \subset H_\varphi$ be a closed subspace satisfying $\pi_\varphi(\mathcal{A})K \subset K$. We show that $K = \{0\}$ or H_φ. This amounts to showing that the projection $P : H_\varphi \longrightarrow K$ is either 0 or the identity operator I on H_φ.

By a remark after Definition 5.7, P commutes with $\pi_\varphi(a)$ for all $a \in \mathcal{A}$. Define a positive linear functional ψ on $\mathcal{A}$ by

$$\psi(x) = \langle \pi_\varphi(x)P\xi_\varphi, P\xi_\varphi \rangle \qquad (x \in \mathcal{A}).$$

We have $\psi \leq \varphi$ since

$$\psi(a^*a) = \|\pi_\varphi(a)P\xi_\varphi\|_\varphi^2 = \|P\pi_\varphi(a)\xi_\varphi\|_\varphi^2 \leq \|\pi_\varphi(a)\xi_\varphi\|_\varphi^2 = \varphi(a^*a).$$

By purity of φ, we have $\psi = \alpha\varphi$ for some $0 \leq \alpha \leq 1$.

For any $a, b \in \mathcal{A}$, we have

$$\langle \alpha\pi_\varphi(a)\xi_\varphi, \pi_\varphi(b)\xi_\varphi \rangle = \alpha\varphi(b^*a) = \psi(b^*a) = \langle \pi_\varphi(a)P\xi_\varphi, \pi_\varphi(b)P\xi_\varphi \rangle$$
$$= \langle P\pi_\varphi(a)\xi_\varphi, P\pi_\varphi(b)\xi_\varphi \rangle = \langle P\pi_\varphi(a)\xi_\varphi, \pi_\varphi(b)\xi_\varphi \rangle.$$

Since $\pi_\varphi(\mathcal{A})\xi_\varphi$ is dense in H_φ, we conclude that $P = \alpha I$ and hence $\alpha = 0$ or 1 since $P^2 = P$. This proves irreducibility of π_φ.

(ii) $\Longrightarrow$ (i) Given that the GNS-representation π_φ is irreducible, we show that φ is pure. Let $0 \leq \psi \leq \varphi$.

For $a, b \in \mathcal{A}$, the Schwarz inequality implies that

$$|\psi(b^*a)|^2 \leq \|\pi_\varphi(a)\xi_\varphi\|^2 \|\pi_\varphi(b)\xi_\varphi\|^2$$

and therefore $\ll \pi_\varphi(a)\xi_\varphi, \pi_\varphi(b)\xi_\varphi \gg := \psi(b^*a)$ defines a positive semidefinite sesquilinear form on the dense subspace $\pi_\varphi(\mathcal{A})\xi_\varphi$ of H_φ. It follows that there is a bounded operator $T \in B(H_\varphi)$ such that

$$\ll \pi_\varphi(a)\xi_\varphi, \pi_\varphi(b)\xi_\varphi \gg = \langle T\pi_\varphi(a)\xi_\varphi, \pi_\varphi(b)\xi_\varphi \rangle \qquad (a, b \in \mathcal{A}).$$

Since

$$\langle T\pi_\varphi(a)\xi_\varphi, \pi_\varphi(a)\xi_\varphi \rangle = \psi(a^*a) \geq 0 \qquad (a \in \mathcal{A})$$

and since $\pi_\varphi(\mathcal{A})\xi_\varphi$ is dense in H_φ, the operator T is a positive operator on H_φ which means $\langle T\eta, \eta \rangle \geq 0$ for all $\eta \in H_\varphi$.

We next show that T commutes with each element in $\pi_\varphi(\mathcal{A})$. For $x, y, z \in \mathcal{A}$, we have

$$\langle T\pi_\varphi(x)\pi_\varphi(y)\xi_\varphi, \pi_\varphi(z)\xi_\varphi \rangle$$
$$= \psi(z^*xy) = \psi((x^*z)^*y) = \langle T\pi_\varphi(y)\xi_\varphi, \pi_\varphi(x^*z)\xi_\varphi \rangle$$
$$= \langle \pi_\varphi(x)T\pi_\varphi(y)\xi_\varphi, \pi_\varphi(z)\xi_\varphi \rangle$$

which implies T commutes with $\pi_\varphi(\mathcal{A})$ by density of $\pi_\varphi(\mathcal{A})$ in H_φ. It follows that all spectral projections of T commutes with each element of $\pi_\varphi(\mathcal{A})$ and they are therefore either 0 or I by irreducibility of π_φ. Hence the spectrum of T reduces to a singleton and $T = \alpha I$ for some $\alpha \geq 0$.

Now we have

$$\psi(a^*a) = \langle T\pi_\varphi(a)\xi_\varphi, \pi_\varphi(a)\xi_\varphi \rangle = \alpha \langle \pi_\varphi(a)\xi_\varphi, \pi_\varphi(a)\xi_\varphi \rangle = \alpha\varphi(a^*a).$$

By Lemma 3.3, we obtain $\psi = \alpha\varphi$. Hence φ is pure. $\qquad\square$

Exercise 5.12. Prove the assertion in *Lemma* 5.1(iv).

Exercise 5.13. Prove the assertion in *Lemma* 5.1(v).

Exercise 5.14. Verify that the GNS-representation π_φ is a $*$-homomorphism.

6. Conditional Expectations

Conditional expectations play an important role in the structure theory of C*-algebras.

A linear map $E : \mathcal{A} \longrightarrow \mathcal{A}$ on a C*-algebra $\mathcal{A}$ is called *contractive* if $\|E(x)\| \leq \|x\|$ for all $x \in \mathcal{A}$. It is called a *projection* if $E^2 = E$ where E^2 is the composite map $E \circ E$. The range $E(\mathcal{A})$ of a contractive projection E is necessarily closed and also $\|E\| = 1$.

Theorem 6.1. *Let $\mathcal{A}$ be a unital C*-algebra and $\mathcal{B}$ be a C*-subalgebra containing the identity of $\mathcal{A}$. Let $E : \mathcal{A} \longrightarrow \mathcal{B}$ be a surjective linear map. The following two statements are equivalent.*

(i) *E is a contractive projection.*
(ii) *E satisfies the conditions*

    (a) *$E(1) = 1$;*
    (b) *$E(x) \geq 0$ for $x \geq 0$;*
    (c) *$E(yxz) = yE(x)z$ for $x \in \mathcal{A}$ and $y, z \in E(\mathcal{A})$.*

Proof. We refer to [17, p. 131] for a proof of (i) $\Rightarrow$ (ii). For (ii) $\Rightarrow$ (i), we observe that conditions (a) and (c) imply $E(y) = y$ for all $y \in E(\mathcal{A})$ and hence $E^2 = E$.

To see that E is contractive, we first note that condition (b) implies $E(x^*) = E(x)^*$ for all $x \in \mathcal{A}$, using the fact that we can write $x = x_1 + ix_2$

where x_1 and x_2 are self-adjoint elements and $x_j = x_j^+ - x_j^-$ with $x_j^\pm \geq 0$, for $j = 1, 2$.

Next, $E((x - E(x))^*(x - E(x)) \geq 0$ gives $E(x^*x) \geq E(x)^*E(x)$ for all $x \in \mathcal{A}$. Given $\|x\| \leq 1$, we have $\|x^*x\| \leq 1$ and $-1 \leq x^*x \leq 1$. By conditions (a) and (b), we have $-1 \leq E(x^*x) \leq 1$. It follows that

$$\|E(x)\|^2 = \|E(x)^*E(x)\| \leq \|E(x^*x)\| \leq 1$$

which proves contractivity of E.   □

In this theorem, it is crucial that the range $E(\mathcal{A}) = \mathcal{B}$ of E is a C*-subalgebra of $\mathcal{A}$. A linear map on a unital C*-algebra $\mathcal{A}$ satisfying condition (ii) in Theorem 6.1 is called a *conditional expectation* of $\mathcal{A}$. Equivalently, a contractive projection E on $\mathcal{A}$ whose range is a C*-subalgebra of $\mathcal{A}$ is a conditional expectation, by Theorem 6.1.

In general, the range $E(\mathcal{A})$ of a contractive projection $E : \mathcal{A} \longrightarrow \mathcal{A}$ need not be a C*-algebra and need not possess all the properties listed in Theorem 6.1. What kind of structure does $E(\mathcal{A})$ have generally? It turns out that $E(\mathcal{A})$ admits a very important triple product which turns it into what is known as a *JB*-triple*.

The concept of a JB*-triple arose originally in the study of infinite-dimensional symmetric Banach manifolds and it is interesting that a close connection to C*-algebras has been found.

A complex Banach space V is called a *JB*-triple* if it is equipped with a continuous triple product $\{\cdot, \cdot, \cdot\} : V^3 \longrightarrow V$, called a *Jordan triple product*, which is symmetric and linear in the outer variables, but conjugate linear in the middle variable, and satisfies

(i) $\{x, y, \{a, b, c\}\} = \{\{x, y, a\}, b, c\} - \{a, \{y, x, b\}, c\} + \{a, b, \{x, y, c\}\}$;

(ii) $\| \exp it(a \square a)\| = 1$ for all $t \in \mathbb{R}$, where $a \square a : V \longrightarrow V$ is defined by $(a \square a)(\cdot) = \{a, a, \cdot\}$;

(iii) $a \square a$ has non-negative spectrum;

(iv) $\|a \square a\| = \|a\|^2$

for $a, b, c, x, y \in V$. A C*-algebra $\mathcal{A}$ is a JB*-triple with the Jordan triple product $\{a, b, c\} := \frac{1}{2}(ab^*c + cb^*a)$ for $a, b, c \in \mathcal{A}$.

We refer to [2] for a proof of the following result as well as the important connections of JB*-triples to geometry and analysis.

Theorem 6.2. *Let $E : \mathcal{A} \longrightarrow \mathcal{A}$ be a contractive projection on a C*-algebra $\mathcal{A}$. Then $E(\mathcal{A})$ is a JB*-triple.*

7. Locally Convex Topologies for $B(H)$

Let H be a complex Hilbert space. The C*-algebra $B(H)$ of bounded operators on H can be equipped with several locally convex topologies. The frequently used topologies are the *operator norm topology*, the *strong operator topology* and the *weak operator topology*, where the word *"operator"* is often omitted.

The operator norm topology is also called the *uniform topology*. It is induced by the operator norm

$$\|T\| = \sup\{\|Tx\| : x \in H, \|x\| \le 1\}.$$

This topology is metrizable and a sequence (T_n) converges to T in this topology if and only if $\lim_{n \to \infty} \|T_n - T\| = 0$.

Both the strong operator topology and the weak operator topology are not metrizable in general. If H is separable, then both topologies are metrizable on the closed unit ball of $B(H)$ (see [17, p. 71] for a proof).

The strong operator topology is the weakest topology for which the mappings

$$T \in B(H) \mapsto Tx \in H \qquad (x \in H)$$

are continuous. It is a locally convex topology and is defined by the seminorms

$$\{p_x : x \in H\}$$

where $p_x(T) = \|Tx\|$ for $x \in H$. A net (T_α) converges to T in the strong operator topology if and only if

$$\|(T_\alpha - T)x\| \longrightarrow 0 \quad \text{as } \alpha \to \infty$$

for each $x \in H$.

The weak operator topology is the weakest topology for which the mappings

$$T \in B(H) \mapsto \langle Tx, y \rangle \in \mathbb{C} \qquad (x, y \in H)$$

are continuous. It is locally convex and is defined by the seminorms

$$\{p_{x,y} : x, y \in H\}$$

where $p_{x,y}(T) = |\langle Tx, y \rangle|$ for $x, y \in H$.

A net (T_α) converges to T in the weak operator topology if and only if

$$\langle (T_\alpha - T)x, y \rangle \longrightarrow 0 \quad \text{as} \quad \alpha \to \infty$$

for every $x, y \in H$. We have the following proper inclusions for the above three topologies:

weak operator topology $\subset$ *strong operator topology* $\subset$ *operator norm topology.*

We have defined positive elements in a C*-algebra. In $B(H)$, an operator T is positive if and only if $\langle Tx, x \rangle \geq 0$ for all $x \in H$. It follows easily from this that, for any $T \in B(H)$, the operator T^*T and TT^* are positive since

$$\langle T^*Tx, x \rangle = \langle Tx, Tx \rangle \geq 0 \quad \text{and} \quad \langle TT^*x, x \rangle = \langle T^*x, T^*x \rangle \geq 0.$$

Also, every projection $P \in B(H)$ is positive. As before, given self-adjoint operators $T, S \in B(H)$, we write $T \geq S$ or $S \leq T$ if $T - S$ is positive. One can define the notion of a *least upper bound* (sup) and a *greatest lower bound* (inf) of a set of self-adjoint operators in the usual way.

Lemma 7.1. *Let $T \in B(H)$ be self-adjoint. Then $T \leq \|T\|\mathbf{1}$.*

Proof. For all $x \in H$, we have $|\langle Tx, x \rangle| \leq \|Tx\|\|x\| \leq \|T\|\|x\|^2 = \langle \|T\|x, x \rangle$. $\square$

Lemma 7.2. *Let $T \geq 0$ in $B(H)$ and let $S \in B(H)$. Then $S^*TS \geq 0$.*

Proof. For each $x \in H$, we have $\langle S^*TSx, x \rangle = \langle TSx, Sx \rangle \geq 0$. $\square$

Lemma 7.3. *Let $S, T \geq 0$ in $B(H)$. If $ST = TS$, then $ST \geq 0$.*

Proof. By spectral theory, there is a positive operator $S^{1/2} \in B(H)$ such that $S = (S^{1/2})^2$ and $S^{1/2}$ is in the C*-subalgebra generated by S and $\mathbf{1}$ (cf. [15, p. 314]). Hence $ST = S^{1/2}S^{1/2}T = S^{1/2}TS^{1/2} \geq 0$. $\square$

Lemma 7.4. *Let $T \in B(H)$ and $0 \leq T \leq P$ for some projection $P \in B(H)$. Then $T = TP = PT$.*

Proof. By Lemma 7.2, we have $((1 - P)T^{1/2})((1 - P)T^{1/2})^* = (1 - P)T(1 - P) = 0$, which gives $(1 - P)T^{1/2} = 0$. Hence $(1 - P)T = 0$ and $T = PT = PT$. $\square$

Proposition 7.5. *Let (T_n) be an increasing sequence of self-adjoint operators in $B(H)$ with $0 \leq T_n \leq 1$. Then the least upper bound $\sup_n T_n$ exists and is the limit of the sequence (T_n) in the strong operator topology.*

Proof. For each $x \in H$, the sequence $(\langle T_n x, x \rangle)$ of real numbers is increasing and bounded above by $\|x\|^2$ and hence

$$Q(x,x) = \sup \langle T_n x, x \rangle = \lim_{n \to \infty} \langle T_n x, x \rangle$$

exists and defines a bounded real quadratic form on H. By polarization, the limit

$$Q(x,y) = \lim_{n \to \infty} \langle T_n x, y \rangle \qquad (x, y \in H)$$

exists and defines a bounded Hermitian form on H. Hence there is a self-adjoint operator $T \in B(H)$ such that $\langle Tx, y \rangle = Q(x,y)$ for $x, y \in H$.

Plainly, (T_n) converges to T in the weak operator topology and $T_n \leq T$. If $S \geq T_n$ for all n, then $\langle Sx, x \rangle \geq \langle T_n x, x \rangle$ for all $x \in H$. Hence $\langle Sx, x \rangle \geq \sup \langle T_n x, x \rangle = \langle Tx, x \rangle$ for each $x \in H$ and $S \geq T$. Therefore $T = \sup_n T_n$.

Finally, we have, for each $x \in H$,

$$\|(T - T_n)x\|^2 \leq \|(T - T_n)^{1/2}\|^2 \|(T - T_n)^{1/2} x\|^2 \leq \langle (T - T_n)x, x \rangle \longrightarrow 0$$

since $0 \leq T - T_n \leq T \leq 1$ implies $\|(T - T_n)^{1/2}\| \leq 1$. This proves (T_n) converges to T in the strong operator topology. $\square$

Corollary 7.6. *Let $T \in B(H)$ and let $E : H \longrightarrow \overline{T(H)}$ be the natural orthogonal projection. Then E is the smallest projection satisfying $ET = T$.*

If T is self-adjoint, then E is the smallest projection satisfying $ET = T = TE$. If $0 \leq T \leq 1$, then $T \leq E$ and E is the limit of a sequence of polynomials in T without constant term, in the strong operator topology.

Proof. Let $P \in B(H)$ be a projection. Then $PT = 0$ if and only if $PE = 0$ since $E(H) = \overline{T(H)}$. In particular, for $P = 1 - E$, we have $(1 - E)T = 0$ which gives $ET = T$. If T is self-adjoint, then we also have $T = TE$.

Given a projection $P \in B(H)$ satisfying $PT = T$, we have $(1 - P)T = 0$ and therefore $(1 - P)E = 0$ and

$$\langle Px, x \rangle = \langle PEx, x \rangle + \langle P(1 - E)x, x \rangle \geq \langle Ex, x \rangle,$$

where $P(1 - E) \geq 0$ since $PE = E = EP$.

Let $0 \leq T \leq 1$. Then $E - T = E(1 - T) \geq 0$ by Lemma 7.3 and hence $E \geq T$.

We note that $\|T\| \leq 1$ and also $T \leq T^{1/2}$ since $T^{1/2} - T = T^{1/4}(1 - T^{1/2})T^{1/4} \geq 0$ where $T^{1/2} \leq \|T^{1/2}\|1 \leq 1$.

Since $T^{1/2}$ is a norm limit of polynomials in T without constant term, we have $ET^{1/2} = T^{1/2} = T^{1/2}E$ and hence $T^{1/2} = ET^{1/2}E \leq E1E = E$. It follows that

$$T \leq T^{1/2} \leq T^{1/4} \leq \cdots \leq T^{1/2^n} \leq \cdots \leq E.$$

Let $Q = \sup\{T^{1/2^n} : n = 1, 2, \ldots\}$. Then $Q \leq E$. By Proposition 7.5, we have $Q = \lim_{n \to \infty} T^{1/2^n}$, where the limit is taken in the strong operator topology and

$$Q^2 = \lim_{n \to \infty} (T^{1/2^n})^2 = \lim_{n \to \infty} T^{1/2^{n-1}} = Q.$$

Now $T \leq Q$ implies $QT = T = TQ$ and hence $E \leq Q$. Therefore $E = Q$ and E is the strong limit of a sequence of polynomials in T without constant term. $\qquad\square$

Definition 7.7. Let $T \in B(H)$. The projection $E : H \longrightarrow \overline{T(H)}$ is called the *range projection* of T and is denoted by $[T]$.

Plainly, the range projection $[P]$ of a projection P is itself. Given $T \in B(H)$, we have

$$\|Tx\|^2 = \langle T^*Tx, x \rangle = \|(T^*T)^{1/2}x\|^2 \qquad (x \in H)$$

and hence T and $(T^*T)^{1/2}$ have the same kernel. Since the orthogonal complement $\overline{T^*(H)}^{\perp}$ of $\overline{T^*(H)}$ is the kernel $T^{-1}(0)$, it follows that $[T^*] = [(T^*T)^{1/2}]$.

8.   von Neumann Algebras

In this final section we discuss a very important class of C*-algebras, namely, the von Neumann algebras. They were first introduced by von Neumann in [11] and studied further by Murray and von Neumann in a series of seminal papers [8–10, 12, 13], under the name of *"rings of operators"*.

We first recall that the algebra $B(H)$ of bounded operators on a Hilbert space H has a predual which is the Banach space $\mathrm{Tr}(H)$ of trace-class operators.

A positive operator $T \in B(H)$ is of *trace class* if

$$\text{trace}(T) := \sum_{\alpha} \langle T\xi_\alpha, \xi_\alpha \rangle < \infty$$

for some orthonormal basis (ξ_α) of H, where the sum does not depend on the choice of (ξ_α). In fact,

$$\sum_{\alpha} \langle T\xi_\alpha, \xi_\alpha \rangle = \sum_{\alpha} \|T^{1/2}\xi_\alpha\|^2 = \sum_{\beta}\sum_{\alpha} |\langle T^{1/2}\xi_\alpha, \eta_\beta \rangle|^2 = \sum_{\beta} \langle T\eta_\beta, \eta_\beta \rangle$$

for any orthonormal basis (η_β) of H. An operator $T \in B(H)$ is of *trace class* if $|T|$ is of trace class as defined above, in which case we define the trace norm $\|T\|_1 = \text{trace}(|T|)$. With this norm, $\text{Tr}(H)$ is a Banach space. It is well known that $\text{Tr}(H)$ is a two-sided ideal in $B(H)$ although it is not closed in the weak operator topology unless $\dim H < \infty$.

We can identify $B(H)$ with the dual $\text{Tr}(H)^*$ via the linear isometry

$$T \in B(H) \mapsto \psi_T \in \text{Tr}(H)^*, \quad \psi_T(S) = \text{trace}(TS) \quad (S \in \text{Tr}(H)).$$

In this duality, the weak* topology of $\text{Tr}(H)^* = B(H)$ coincides with the weak operator topology of $B(H)$, and a subspace X of $B(H) = \text{Tr}(H)^*$ is weak* closed if and only if it is the dual of a quotient space of $\text{Tr}(H)$, namely, $\text{Tr}(H)/X^0$ where $X^0 = \{S \in \text{Tr}(H) : \psi(S) = 0, \forall \psi \in X\}$.

Definition 8.1. Let H be a Hilbert space. A unital C*-subalgebra $\mathcal{A}$ of the algebra $B(H)$ of bounded operators on H is called a *von Neumann algebra* (*acting on H*) if it is closed in the weak operator topology of $B(H)$.

We often omit mentioning the underlying Hilbert space H for a von Neumann algebra $\mathcal{A} \subset B(H)$ if it is understood.

A unital C*-algebra $\mathcal{A}$ is called a *W*-algebra* if it admits a faithful representation $\pi : \mathcal{A} \longrightarrow B(H)$ such that $\pi(\mathcal{A})$ is a von Neumann algebra.

By Corollary 7.6, a von Neumann algebra contains the range projections of its positive elements. Although von Neumann algebras are C*-algebras, they have a distinctive feature characterised by the existence of sufficiently many projections determining their intrinsic structures. For this reason, the study of projections is central in the theory of von Neumann algebras. On the other hand, C*-algebras do not always have non-trivial projections, for instance, the algebra $C[0,1]$ of continuous functions on $[0,1]$, but the representation theory plays an important role in C*-algebras.

Given a self-adjoint operator $T \in B(H)$, we have $|T| = (T^2)^{1/2}$ by functional calculus in Section 3. For any $T \in B(H)$, we define

$$|T| = (T^*T)^{1/2}.$$

The spectrum $\sigma(|T|)$ of $|T|$ is contained in $[0, \infty)$ because

$$\sigma(|T|) = \sigma((T^*T)^{1/2})^2) = \{\alpha^2 : \alpha \in \sigma((T^*T)^{1/2})\}.$$

By a remark following Definition 7.7, we have $[T^*] = [|T|]$.

Lemma 8.2. *Let $T \in B(H)$ and let*

$$T_n = \left(\frac{1}{n} + |T|\right)^{-1} |T| \qquad (n = 1, 2, \ldots).$$

Then $T_n \leq 1$ and the sequence (T_n) is increasing.

Proof. This follows from functional calculus by considering the C*-subalgebra of $B(H)$ generated by T^*T and $\mathbf{1}$. □

By Proposition 7.5, the sequence (T_n) above converges to some $S \in B(H)$ in the strong operator topology. It follows that the sequence $(T_n - T_m)^2$ converges to 0 in the strong, and hence, weak operator topology since for each $x \in H$, we have

$$\|(T_n - T_m)^2 x\| \leq \|T_n - T_m\| \|(T_n - T_m)x\|$$

$$\leq 2\|(T_n - T_m)x\| \longrightarrow 0 \quad \text{as } n, m \to \infty.$$

We now show that every element T in a von Neumann algebra admits a *polar decomposition* analogous to the polar decomposition of a complex number $z = e^{i\theta}|z|$.

Proposition 8.3. *Let $\mathcal{M} \subset B(H)$ be a von Neumann algebra and let $T \in \mathcal{M}$. Then $T = U|T|$ for some $U \in \mathcal{M}$ such that U^*U is the range projection of $|T|$.*

Proof. We have $|T| \in \mathcal{M}$ and $[|T|] \in \mathcal{M}$. For $n = 1, 2, \ldots$, let

$$U_n = T \left(\frac{1}{n} + |T|\right)^{-1} \in \mathcal{M}.$$

Since the range projection $[|T|]$ commutes with $(\frac{1}{n} + |T|)^{-1}$ and $T = T[|T|]$, we have $U_n = U_n[|T|]$ and

$$(U_n - U_m)^*(U_n - U_m) = (T_n - T_m)^2$$

with T_n defined in Lemma 8.2. Hence

$$\|(U_n - U_m)x\|^2 = \langle (T_n - T_m)^2 x, x \rangle \longrightarrow 0 \quad \text{as } n, m \to \infty.$$

It follows that the sequence (U_n) converges to some $U \in \mathcal{M}$ in the strong operator topology and $U = U[|T|]$.

Since $U_n|T| = T\left(\frac{1}{n} + |T|\right)^{-1}|T|$ converges to T in the strong operator topology, we have $T = U|T|$.

Finally, we have $U^*U = |T|U^*U|T|$ which gives $(|T|U^*U - |T|)|T| = 0$ and hence $(|T|U^*U - |T|)[|T|] = 0$. It follows that $|T|U^*U[|T|] - |T| = 0$ and $[|T|]U^*U|T| - |T| = 0$. Therefore $([|T|]U^*U - 1)[|T|] = 0$ and $[|T|]U^*U[|T|] = [|T|]$, giving $U^*U = [|T|]$ since $U = U[|T|]$. $\qquad\square$

Definition 8.4. An element u in a C*-algebra $\mathcal{A}$ is called a *partial isometry* if u^*u is a projection.

A partial isometry $u \in \mathcal{A} \subset B(H)$ has the polar decomposition

$$u = uu^*u$$

since $|u| = u^*u$. The self-adjoint element uu^* is also a projection since $\sigma(uu^*) \cup \{0\} = \sigma(u^*u) \cup \{0\} = \{1, 0\}$. We call u^*u the *initial projection* of u, and uu^* the *final projection*.

Two projections $p, q \in \mathcal{M}$ are said to be *equivalent*, in symbol $p \sim q$, if there is a partial isometry $v \in \mathcal{M}$ such that $p = v^*v$ and $q = vv^*$. If $p \leq q$, we say that q *contains* p. If $p \sim z \leq q$ for some subprojection z of q, we write $p \preceq q$.

We now classify von Neumann algebras using projections. As usual, the centre $\mathcal{Z}$ of a C*-algebra $\mathcal{A}$ is the subalgebra of $\mathcal{A}$ consisting of elements which commute with every element in $\mathcal{A}$. Given a von Neumann algebra $\mathcal{M} \subset B(H)$, we define its *commutant* $\mathcal{M}'$ by

$$\mathcal{M}' = \{T \in B(H) : TS = ST \, \forall S \in \mathcal{M}\}.$$

The centre of $\mathcal{M}$ is $\mathcal{M} \cap \mathcal{M}'$. If the centre of $\mathcal{M}$ is trivial, that is, if the centre consists of only scalar multiples of the identity, then $\mathcal{M}$ is called a *factor*.

A projection in the centre of $\mathcal{M}$ is called a *central projection*. The identity is the only non-zero central projection in $B(H)$.

Given a projection p in a von Neumann algebra $\mathcal{M}$, it is evident that the reduced algebra $p\mathcal{M}p$ is also a von Neumann algebra.

Definition 8.5. A projection p in a von Neumann algebra $\mathcal{M}$ is called *abelian* if the algebra $p\mathcal{M}p$ is commutative.

Definition 8.6. A von Neumann algebra $\mathcal{M}$ is said to be of *type I* if every non-zero central projection $p \in \mathcal{M}$ contains a non-zero abelian projection.

Evidently every abelian von Neumann algebra is of type I. Also, $B(H)$ is of type I since every projection $p \in B(H)$ with $\dim p(H) = 1$ is abelian as $pB(H)p = \mathbb{C}p$.

Example 8.7. Let $\mathcal{M}$ be a von Neumann algebra. A non-zero projection $p \in \mathcal{M}$ is called *minimal* if $p\mathcal{M}p = \mathbb{C}p$. Trivially minimal projections are abelian. If $\mathcal{M}$ is a factor and contains a minimal projection, it must be of type I.

Let $\dim \mathcal{M} < \infty$. We may assume $\mathcal{M} \subset B(H)$ with $\dim H < \infty$ by the universal representation of $\mathcal{M}$. Then $\dim p(H) < \infty$ for every $p \in \mathcal{M}$ and, by a simple dimension argument, $\mathcal{M}$ contains a non-zero projection p such that $0 \neq q \leq p \Rightarrow q = p$ for any projection $q \in \mathcal{M}$. We must have $p\mathcal{M}p = \mathbb{C}p$, that is, p is a minimal projection in $\mathcal{M}$. Indeed, the condition on p implies that every non-zero element in $p\mathcal{M}p$ has range projection p. If $T \in p\mathcal{M}p$ and if $\alpha p - T \neq 0$ for some $\alpha \in \mathbb{C}\backslash\{0\}$, then $(\alpha p - T)^{-1}(0) = p^{-1}(0)$ and hence α is not an eigenvalue of T since $(\alpha 1 - T)(x) = 0$ for $x \in H$ implies $(\alpha p - T)(x) = 0$, giving $p(x) = 0$ and

$$\alpha(1 - p)(x) = (\alpha - T)(1 - p)(x) = 0.$$

Hence, if T has a non-zero eigenvalue α, then $T = \alpha p$.

It follows that $\mathcal{M}$ is of type I since $z\mathcal{M}z$ is finite dimensional and contains a minimal projection for every central projection $z \in \mathcal{M}$.

Definition 8.8. A projection p in a von Neumann algebra $\mathcal{M}$ is called *finite* if $p = q$ for any projection q satisfying $p \sim q \leq p$.

In other words, a finite projection is one which is not equivalent to any of its proper subprojections, in analogy to the concept of a finite set. A von Neumann algebra $\mathcal{M}$ is called *finite* if the identity $\mathbf{1}$ is a finite projection.

A finite type I von Neumann algebra is said to be of *type I_f*.

Proposition 8.9. *In a von Neumann algebra $\mathcal{M}$, every abelian projection is finite.*

Proof.   Let p be abelian and let $p \sim q \leq p$. Then there is a partial isometry $v \in \mathcal{M}$ such that $p = v^*v$ and $q = vv^*$. We have $pvp = pvv^*v = pqv = qv = vv^*v = v$ and hence v is in the abelian algebra $p\mathcal{M}p$. Therefore $p = vv^* = q$. $\square$

Example 8.10. On any Hilbert space H, the *rank* of an operator $T \in B(H)$ is defined to be the dimension $\dim T(H)$ of its range. Two projections p and q are equivalent in $B(H)$ if and only if they have the same rank, in which case the partial isometry implementing the equivalence is the natural extension of the isometry between $p(H)$ and $q(H)$. It follows that the finite projections in $B(H)$ are exactly the finite rank projections. In particular, every finite-dimensional von Neumann algebra is of type I_f.

Example 8.11. In contrast to the case of the full algebra $B(H)$, a finite projection p in a von Neumann algebra $\mathcal{M}$ need not have finite rank. Let $\ell_2(\mathbb{N})$ be the Hilbert space of square-summable sequences. An operator $T \in B(\ell_2(\mathbb{N}))$ can be represented as an infinite matrix (a_{ij}) with $a_{ij} = \langle Te_j, e_i \rangle$, where $\{e_1, e_2, \ldots\}$ is the standard basis in $\ell_2(\mathbb{N})$, namely, e_i is the sequence whose terms are 0 except the ith term which is 1.

Let $\mathcal{M} \subset B(\ell_2(\mathbb{N}))$ be the abelian von Neumann subalgebra consisting of the diagonal matrices. The $\mathcal{M}$ contains the projection

$$
p = \begin{pmatrix} 0 & & & \\ & 1 & & \\ & & 1 & \\ & & & \ddots \end{pmatrix} : (x_1, x_2, x_3, \cdots) \in \ell_2(\mathbb{N}) \mapsto (0, x_2, x_3, \cdots) \in \ell_2(\mathbb{N})
$$

which is a finite projection in $\mathcal{M}$, but has infinite rank and is not a finite projection in $B(\ell_2(\mathbb{N}))$.

Definition 8.12. A von Neumann algebra $\mathcal{M}$ is said to be of *type II* if it has no non-zero abelian projection and every non-zero central projection in $\mathcal{M}$ contains a non-zero finite projection. A finite type II von Neumann algebra is said to be of *type II$_1$*.

Definition 8.13. A von Neumann algebra $\mathcal{M}$ is said to be of *type III* if it contains no non-zero finite projection.

Definition 8.14. A von Neumann algebra is said to be *properly infinite* if it contains no non-zero finite central projection.

Definition 8.15. A properly infinite type I von Neumann algebra is said to be of type I_∞. A properly infinite type II von Neumann algebra is said to be of *type II$_\infty$*.

Theorem 8.16. *A von Neumann algebra* $\mathcal{M}$ *decomposes uniquely into five direct summands*:

$$\mathcal{M} = \bigoplus_j \mathcal{M}_j,$$

where $\mathcal{M}_j$ *is either* $\{0\}$ *or of type* j, *for* $j = \mathrm{I}_f, \mathrm{I}_\infty, \mathrm{II}_1, \mathrm{II}_\infty, \mathrm{III}$.
A factor has one and only one of the above types.

We omit the proof of the above theorem which can be found in many books on operator algebras, for instance, [7, p. 422; 14, p. 174; 16, p. 86; 17, p. 296; 18, p. 25]. The classification of types in Murray and von Neumann's original paper [8] is defined in terms of a dimension function. It is of interest to note that type III factors have been further classified into types III_λ ($0 \leq \lambda \leq 1$) by Connes [3], who was awarded a Fields Medal in 1983 for this remarkable work, among his other contribution.

We leave out the discussion of two celebrated density theorems in operator algebras, namely, *Kaplansky's density theorem* and *von Neumann's double commutant theorem* which can also be found in the books mentioned above.

Using the Kaplansky's density theorem, one can show that the universal representation $\pi : \mathcal{A} \longrightarrow B(H)$ of a C*-algebra $\mathcal{A}$ can be extended to a linear (surjective) isometry $\widetilde{\pi} : \mathcal{A}^{**} \longrightarrow \overline{\pi(\mathcal{A})} \subset B(H)$ where $\overline{\pi(\mathcal{A})}$ is the weak closure of the C*-algebra $\pi(\mathcal{A})$ (see, for example, [16, 1.17.2]). It follows that the second dual $\mathcal{A}^{**}$ carries the von Neumann algebraic structure of $\overline{\pi(\mathcal{A})}$ and $\mathcal{A}$ embeds in $\mathcal{A}^{**}$ as a C*-subalgebra. Further, it can be shown that a C*-algebra $\mathcal{A}$ is a W*-algebra if and only if $\mathcal{A} = N^*$ for some Banach space N (cf. [17, p. 133]).

Exercise 8.17. Show that $(U_n - U_m)^*(U_n - U_m) = (T_n - T_m)^2$ in Proposition 8.3.

References

[1] W. Arveson, *An Invitation to C*-algebras*. Springer-Verlag, Berlin (1976).
[2] C.-H. Chu, *Jordan Structures in Geometry and Analysis*. Cambridge Tracts in Mathematics, Vol. 190. Cambridge University Press, Cambridge (2012).
[3] A. Connes, Une classification des facteurs de type III. *Ann. Sci. Ecole Norm. Sup.* **6**, 133–252 (1973).
[4] J. Dixmier, *Les C*-algèbres et leurs représentations*. Gauthier-Villars, Paris (1969).

[5] K. R. Goodearl, *Notes on Real and Complex C*-algebras*. Shiva Mathematics Series, Vol. 5. Shiva Publishing Ltd., Nantwich (1982).
[6] R. V. Kadison, Isometries of operator algebras. *Ann. of Math.* **54**, 325–338 (1951).
[7] R. V. Kadison and J. R. Ringrose, *Fundamentals of the Theory of Operator Algebras*. Academic Press, London (1983, 1986).
[8] F. J. Murray and J. von Neumann, On rings of operators. *Ann. of Math.* **37**, 116–229 (1936).
[9] F. J. Murray and J. von Neumann, On rings of operators II. *Trans. Amer. Math. Soc.* **41**, 208–248 (1937).
[10] F. J. Murray and J. von Neumann, On rings of operators IV. *Ann. of Math.* **44**, 716–808 (1943).
[11] J. von Neumann, Zur Algebra der Funktionaloperationen und Theorie der normalen Operatoren. *Math. Ann.* **102**, 370–427 (1930).
[12] J. von Neumann, On rings of operators III. *Ann. of Math.* **41**, 94–161 (1940).
[13] J. von Neumann, On rings of operators. Reduction theory. *Ann. of Math.* **50**, 401–485 (1949).
[14] G. K. Pedersen, *C*-algebras and Their Automorphism Groups*. Academic Press, London (1979).
[15] W. Rudin, *Functional Analysis*. McGraw-Hill, New York (1973).
[16] S. Sakai, *C*-algebras and W*-algebras*. Springer-Verlag, Berlin (1971).
[17] M. Takesaki, *Theory of Operator Algebras I*. Springer-Verlag, Berlin (1979).
[18] D. M. Topping, *Lectures on von Neumann Algebras*. Van Nostrand, London (1971).

Solutions of selected exercises

Exercise 2.15. $(1 - a)^{-1} = \sum_{n=0}^{\infty} a^n$.

Exercise 2.16. If there exists $x \in \mathcal{A} \backslash \mathbb{C}\mathbf{1}$, then $\sigma(x) = \emptyset$ (why?) which is impossible.

Exercise 4.5. If φ is a *surjective* isometry, then it is a well-known result of Kadison [6] that $\varphi(xx^*x) = \varphi(x)\varphi(x)^*\varphi(x)$ for all $x \in \mathcal{A}$. In particular, if $\mathcal{A}$ and $\mathcal{B}$ are unital and $\varphi(\mathbf{1}) = \mathbf{1}$, then φ is a $*$-isomorphism.

Exercise 4.6. If φ is not surjective, the answer is negative, but one can show that it is *"almost"* a $*$-homomorphism *locally*, meaning: for each $a \in \mathcal{A}$, we have

$$\varphi(xx^*x) = \varphi(x)\varphi(x)^*\varphi(x) \quad (x \in C(a))$$

modulo a projection $p \in \mathcal{B}^{**}$, that is, both sides of the above equality should be multiplied on the right by p. This result and more details can be found in [2, §3.4].

Exercise 5.12. The linear functional $\varphi_y : x \in \mathcal{A} \mapsto \varphi(y^*xy) \in \mathbb{C}$ is positive and hence $\|\varphi_y\| = \varphi_y(1)$ by (iii).

Exercise 5.13. $\varphi + \psi$ is positive and hence $\|\varphi + \psi\| = (\varphi + \psi)(1) = \varphi(1) + \psi(1)$.

Exercise 8.17. Observe that the left-hand side equals

$$\left(\left(\frac{1}{n} + |T| \right)^{-1} - \left(\frac{1}{m} + |T| \right)^{-1} \right) T^*T \left(\left(\frac{1}{n} + |T| \right)^{-1} - \left(\frac{1}{m} + |T| \right)^{-1} \right)$$

and use $T^*T = |T|^2$.

Chapter 4

Special Functions

Rod Halburd

Department of Mathematics, University College London
Gower Street, London WC1E 6BT, UK
r.halburd@ucl.ac.uk

Special functions are natural generalisations of the elementary functions. We will concentrate on two classes of special functions: elliptic
functions and functions defined as solutions of special differential equations. Weierstrass and Jacobi elliptic functions will be described and a
number of identities will be derived, including differential equations and
addition laws. The Riemann equation is the most general second-order
homogeneous linear equation with three regular singular points in the
extended complex plane. Its solutions can be mapped to the hypergeometric equation. We will discuss the analytic continuation of solutions.
Along the way we will meet Kummer's solutions and the monodromy
group. For matrix equations with regular singular points, we will derive
the Schlesinger equations as the condition that the monodromy remains
fixed as the locations of the singular points vary. The simplest non-trivial
case is equivalent to the sixth Painlevé equation. The Schlesinger and
Painlevé equations are important integrable equations.

1. Introduction

This chapter is based on material covered in part of my London Taught
Course Centre course on Complex Analysis. The other half concerned the
value distribution of meromorphic functions. Special functions are natural generalisations of elementary functions. They are usually considered
"special" because they satisfy a number of identities, including differential
or difference equations. They also have simple singularity structure when
considered in the complex domain.

Elliptic functions are natural generalisations of trigonometric functions. They appear in many applications such as the equations of motion for a simple pendulum and in the calculation of the perimeters of several simple geometric figures including ellipses (hence the name "elliptic") and lemniscates. We will describe elliptic functions in Section 2. We will concentrate on the complex analytic theory of these functions as we will not have space to describe, e.g. their importance in number theory.

In Section 3 we will study Fuchsian linear differential equations. We will mostly be concerned with the hypergeometric equation. Various limits of this equation in which singularities merge include many of the important classical equations of mathematical physics: the confluent hypergeometric, Bessel, Hermite and Airy equations. The general solution of the hypergeometric equation is branched at 0, 1 and ∞. We will describe the analytic continuation of solutions and construct the monodromy matrices that describe how a basis of solutions changes as it is continued along closed paths. In Section 4 we will study matrix Fuchsian equations. We will show that demanding that the monodromy data remains fixed as we vary the locations of the singularities results in the Schlesinger equations. These are important integrable nonlinear equations. In the case of four regular singularities, the resulting Schlesinger equation is equivalent to the sixth Painlevé equation.

2. Elliptic Functions

Two fundamental properties of trigonometric functions are that they are meromorphic and periodic. An elliptic function is a meromorphic function that has two independent periods.

Definition 2.1. Let ω_1 and ω_2 be two complex numbers whose ratio is not real. Then any function f for which

$$f(z + 2\omega_1) = f(z) \quad \text{and} \quad f(z + 2\omega_2) = f(z)$$

for all z in the domain of f is said to be *doubly periodic* with periods $2\omega_1$ and $2\omega_2$. If f is doubly periodic and meromorphic (in all $\mathbb{C}$) then it is said to be an *elliptic function*.

A *period parallelogram* with respect to the periods $2\omega_1$ and $2\omega_2$ is a region of the form $\{z = z_0 + 2(s\omega_1 + t\omega_2) : 0 \leq s < 1, 0 \leq t < 1\}$, for some

$z_0 \in \mathbb{C}$. Because of its double periodicity, if we know an elliptic function on one of its period parallelograms, we know it in the whole complex plane, since f has the same value at z_1 and $z_2 = z_1 + 2(m\omega_1 + n\omega_2)$, where m and n are integers. The points z_1 and z_2 are said to be *congruent* with respect to the periods $2\omega_1$ and $2\omega_2$.

We will begin by deriving some simple properties of elliptic functions.

Theorem 2.2 (Liouville). *Any entire elliptic function is a constant.*

Proof. If f is entire it must be bounded on the closure of any period-parallelogram. From the double periodicity of f it follows that f is bounded on $\mathbb{C}$. All bounded entire functions are constants (Liouville's theorem).
□

Theorem 2.3. *The sum of the residues of any elliptic function in any period parallelogram is zero.*

Proof. Let P be a period parallelogram for the elliptic function f. The sum of the residues of f in P is given by

$$\frac{1}{2\pi i} \int_C f(z)\, dz,$$

where C is the boundary of P traced in the positive direction. Note that f has the same values at congruent points on opposite sides of P, however, the direction of integration is opposite on opposite sides of P. Hence the above integral vanishes.
□

2.1. The Weierstrass $\wp$ function

The Weierstrass $\wp$ function is defined by

$$\wp(z) := \frac{1}{z^2} + \sideset{}{'}\sum_{m,n} \left\{ \frac{1}{(z - \Omega_{mn})^2} - \frac{1}{\Omega_{mn}^2} \right\}, \tag{2.1}$$

where $\Omega_{mn} = 2(m\omega_1 + n\omega_2)$ and $\sum'_{m,n}$ denotes the sum over all integer m and n excluding $(m,n) = (0,0)$. The series (2.1) converges uniformly on any compact set not containing Ω_{mn} for all integers m and n. The Weierstrass $\wp$ function is even and analytic everywhere in the complex plane except at the points $z = \Omega_{mn}$, where it has double poles.

Theorem 2.4. *The Weierstrass $\wp$ function is an elliptic function with periods $2\omega_1$ and $2\omega_2$.*

Proof. Since the series (2.1) converges uniformly on all compact sets avoiding the poles at $z = \Omega_{mn}$, we can differentiate the series term-by-term to obtain the uniformly convergent series

$$\wp'(z) = -2 \sum_{m,n=-\infty}^{\infty} \frac{1}{(z - 2[m\omega_1 + n\omega_2])^3}. \tag{2.2}$$

Hence $\wp'$ is meromorphic. On replacing z with $z + 2\omega_j$, $j = 1, 2$, in equation (2.2) and incrementing either the m or the n index, we see that

$$\wp'(z + 2\omega_j) = \wp'(z), \quad j = 1, 2, \tag{2.3}$$

for all $z \in \mathbb{C}$. Therefore $\wp'$ is a doubly periodic meromorphic function — i.e. it is elliptic. Integrating equation (2.3) with respect to z yields

$$\wp(z + 2\omega_j) = \wp(z) + \kappa_j, \tag{2.4}$$

for some constant κ_j. Substituting $z = -\omega_j$ into equation (2.4) and using the fact that $\wp$ is even gives $\kappa_j = 0$, $j = 1, 2$. Hence $\wp$ has periods $2\omega_1$ and $2\omega_2$. $\qquad\square$

We mention without proof that any even elliptic function can be expressed as a rational function of the Weierstrass $\wp$ function with the same periods. A corollary of this is the fact that any elliptic function f can be expressed as $Q(\wp(z)) + R(\wp(z))\wp'(z)$, where f and $\wp$ have the same periods and Q and R are rational with respect to $\wp$.

2.2.   An ODE for $\wp$

Note that $\wp'$ is an elliptic function and so is any polynomial P of $\wp$ and $\wp'$. The only place where $P(\wp, \wp')$ can have a pole is at Ω_{mn}. We will now construct a polynomial P such that $P(\wp, \wp')$ has no pole at the origin. It is therefore entire and hence constant by Liouville's theorem.

Note that since $\wp$ is even, its Laurent expansion about $z = 0$ contains only even powers of z. From equation (2.1) we have for $z \to 0$

$$\wp(z) = \frac{1}{z^2} + 3G_2 z^2 + 5G_3 z^4 + \cdots,$$

where $G_2 := \sum'_{m,n} \Omega_{mn}^{-4}$, $G_3 := \sum'_{m,n} \Omega_{mn}^{-6}$. So

$$\wp' = -\frac{2}{z^3} + 6G_2 z + 20G_3 z^4 + \cdots .$$

In order to cancel the leading terms in the series for $\wp$ and $\wp'$ it is natural to consider

$$\wp'^2 - 4\wp^3 = -60\frac{G_2}{z^2} - 140G_3 + O(z^2) = -60G_2\wp - 140G_4 + O(z^2).$$

It follows that $f := \wp'^2 - 4\wp^3 + 60G_2\wp + 140G_4$ is an elliptic function with no pole in any fundamental period parallelogram containing $z = 0$ and hence it is entire. It follows that f is a constant. As $z \to 0$, $f = O(z^2)$, so $f = 0$. Therefore $\wp$ satisfies the ODE

$$\wp'^2 = 4\wp^3 - g_2\wp - g_3, \tag{2.5}$$

where the constants g_2 and g_3 are given by

$$g_2 := 60 \sum_{m,n}' \Omega_{mn}^{-4}, \quad g_3 := 140 \sum_{m,n}' \Omega_{mn}^{-6}. \tag{2.6}$$

Conversely, the general solution of

$$\left(\frac{dy}{dz}\right)^2 = 4y^3 - g_2y - g_3,$$

is given by $y(z) = \wp(z - z_0)$, where z_0 is an arbitrary constant and the Ω_{mn} satisfy (2.6).

2.3. The addition law for $\wp$

We begin by proving the following result.

Theorem 2.5. *Let f be an elliptic function with fundamental period parallelogram P. Let $a_1, \ldots, a_m$ and $b_1, \ldots, b_n$ denote the zeros and poles of f respectively in P, counted according to multiplicity (i.e. if f has a double zero at $z = z_0$, then z_0 appears twice in the list). Then*

$$\sum_{i=1}^{m} a_i - \sum_{j=1}^{n} b_j$$

is a period of f.

Proof. Let $P = \{z \in \mathbb{C} : z = c + 2(s\omega_1 + t\omega_2), 0 \le s < 1 \text{ and } 0 \le t < 1\}$. Now

$$2\pi i \left(\sum_{i=1}^{m} a_i - \sum_{j=1}^{n} b_j \right)$$

$$= \int_{\partial P} z \frac{f'(z)}{f(z)} dz$$

$$= \left(\int_c^{c+2\omega_1} + \int_{c+2\omega_1}^{c+2\omega_1+2\omega_2} + \int_{c+2\omega_1+2\omega_2}^{c+2\omega_2} + \int_{c+2\omega_2}^{c} \right) z \frac{f'(z)}{f(z)} dz$$

$$= \left(\int_c^{c+2\omega_1} - \int_{c+2\omega_2}^{c+2\omega_1+2\omega_2} + \int_{c+2\omega_1}^{c+2\omega_1+2\omega_2} - \int_c^{c+2\omega_2} \right) z \frac{f'(z)}{f(z)} dz$$

$$= \int_c^{c+2\omega_1} \left(z \frac{f'(z)}{f(z)} - (z + 2\omega_2) \frac{f'(z + 2\omega_2)}{f(z + 2\omega_2)} \right) dz$$

$$+ \int_c^{c+2\omega_2} \left((z + 2\omega_1) \frac{f'(z + 2\omega_1)}{f(z + 2\omega_1)} - z \frac{f'(z)}{f(z)} \right) dz$$

$$= -2\omega_2 \int_c^{c+2\omega_1} \frac{f'(z)}{f(z)} dz + 2\omega_1 \int_c^{c+2\omega_2} \frac{f'(z)}{f(z)} dz,$$

where we have used the $2\omega_j$-periodicity of f to obtain the last line. Finally note that both integrals give $\log f$ evaluated at points differing by a period of f. Due to the branching of log, these factors are both of the form $2\pi i$ times an integer. $\qquad\square$

Let f be an elliptic function with fundamental period parallelogram P. We will show that f takes every value the same number of times in P (counting multiplicities.) For any $\alpha \in \mathbb{C}$, the number of times $f(z) = \alpha$ minus the number of poles of f in P is given by

$$\frac{1}{2\pi i} \oint_{\partial P} \frac{f'(z)}{f(z) - \alpha} dz.$$

Note that the integrand, g, is an elliptic function with the same periods as f. Hence, this integral vanishes because g has the same values at congruent points on opposite sides of the parallelogram but the direction of integration is the opposite.

Let $\wp$ be the Weierstrass elliptic function with fundamental periods $2\omega_1$ and $2\omega_2$ and let $\omega_3 = -\omega_1 - \omega_2$. Since $\wp$ is even, $\wp'$ is odd. Hence $\wp'(\omega_i) = -\wp'(-\omega_i) = -\wp'(-\omega_i + 2\omega_i)$ (since $\wp'$ has period $2\omega_i$). So $\wp'(\omega_i) = 0$.

Let $e_i = \wp(\omega_i)$, $i = 1, 2, 3$. We will now show that the e_i's are distinct and are the three roots of the equation $4t^3 - g_2 t - g_3 = 0$. Recall that the only pole of $\wp$ in P is a double pole at the origin. Therefore, $\wp'$ has exactly one pole (a triple pole) in P. So counting multiplicities, $\wp'$ must vanish exactly three times in P. We have already found three such points, namely ω_1, ω_2 and $-\omega_3$. Also, since $\wp$ takes every value twice in P (counting multiplicities), and $\wp'(\omega_i) = 0$, then ω_i is the only point in P where $\wp(z) = e_i$. Therefore all the e_i's are distinct so from equation (2.5),

$$\wp'^2 = 4\wp^3 - g_2\wp - g_3 = 4(\wp - e_1)(\wp - e_2)(\wp - e_3). \tag{2.7}$$

Next we will derive the addition law for $\wp$. Let z_1 and z_2 be two points where $\wp$ is analytic and where $\wp(z_1) \neq \wp(z_2)$ (i.e. $z_1 - z_2$ is not a period or a half-period of $\wp$). Define the elliptic function

$$f(z) := c_1 + c_2\wp(z) - \wp'(z),$$

where the constants c_1 and c_2 are determined by the condition

$$\wp'(z_j) = c_1 + c_2\wp(z_j), \quad j = 1, 2.$$

Note that f has only one pole in the fundamental period parallelogram P, namely a triple pole at $z = 0$. Hence f vanishes exactly three times in P, counting multiplicities. Two zeros are z_1 and z_2. We also know from Theorem 2.5 that the sum of the location of the three zeros must be congruent to 0. Hence the third zero must be congruent to $-z_1 - z_2$. Hence $-z_1 - z_2$ must be a zero of f and so

$$c_1 + c_2\wp(-z_1 - z_2) - \wp'(-z_1 - z_2) = 0.$$

Eliminating c_1 and c_2 from the above gives

$$\begin{vmatrix} \wp(z_1) & \wp'(z_1) & 1 \\ \wp(z_2) & \wp'(z_2) & 1 \\ \wp(z_1 + z_2) & -\wp'(z_1 + z_2) & 1 \end{vmatrix} = 0.$$

This is called the addition law for $\wp$. Explicitly it gives

$$\wp(z_1 + z_2) = \frac{1}{4}\left\{\frac{\wp'(z_1) - \wp'(z_2)}{\wp(z_1) - \wp(z_2)}\right\}^2 - \wp(z_1) - \wp(z_2). \tag{2.8}$$

2.4.   *Elliptic integrals of the first kind*

We have seen that an integral involving the square root of a cubic function can be evaluated in terms of the (inverse) Weierstrass elliptic function. In this section we will consider integrals involving the square root of quartic functions and we will see that they too can be solved in terms of elliptic functions.

Jacobi's (incomplete) integral of the first kind is

$$F(\zeta; k) := \int_0^\zeta [(1 - z^2)(1 - k^2 z^2)]^{-1/2} dz, \qquad (2.9)$$

where k is a constant called the modulus, which in general we take not to be 1 or -1. Clearly F is branched — its value at any point depends on the path on integration, due to the square-root singularities of the integrand at $z = \pm 1$ and $z = \pm 1/k$.

Note that $F(\zeta, 0) = \sin^{-1} \zeta$, which is the inverse of an entire periodic function. We will see that in general $F(\zeta, k)$ is the inverse of an elliptic function, which is denoted by $\text{sn}(\zeta; k)$.

We will begin by considering the behaviour of the mapping $\zeta \mapsto F(\zeta; k)$ in the complex plane. Consider the case in which k is real and $0 < k < 1$. We will first determine the image of the real axis under this transformation where the path of integration is along the real axis except at the singularities $z = \pm 1$, $z = \pm 1/k$ where the path of integration is taken to follow small semi-circles lying in the upper half-plane. With this choice for the path of integration, $F(0, k) = 0$ and $F(1, k) = K$, where K is the positive real number

$$K \equiv K(k) = \int_0^1 [(1 - z^2)(1 - k^2 z^2)]^{-1/2} dz,$$

called the *complete elliptic integral of the first kind*. So under the mapping $\zeta \mapsto F(\zeta; k)$, the image of the segment of real axis from 0 to 1 is the segment of the real axis from 0 to K. Now as z passes the value $z = 1$ along a small semi-circle $z = 1 - \epsilon e^{-i\theta}$ as θ varies from 0 to π, the integrand becomes a pure imaginary number. This gives

$$F(\zeta; k) = K + i \int_1^\zeta [(z^2 - 1)(1 - k^2 z^2)]^{-1/2} dz, \quad 1 < \zeta < 1/k.$$

So the image of the segment of the real axis from 1 to $1/k$ is the straight line segment from K to $K + iK'$, where

$$K' = \int_1^{1/k} [(z^2 - 1)(1 - k^2 z^2)]^{-1/2} dz,$$

is known as the *complementary integral of the first kind*. A change of variables shows that $K' = K(k')$, where $k' := \sqrt{1 - k^2}$ is called the *complementary modulus*.

As z continues past $z = 1/k$ by following a small semi-circle in the upper half-plane and then continuing along the positive real axis, the integrand picks up another factor of i, giving

$$F(\zeta; k) = K + iK' - \int_{1/k}^{\zeta} [(z^2 - 1)(k^2 z^2 - 1)]^{-1/2} dz, \quad \zeta > 1/k.$$

As $\zeta \to \infty$ along the positive real axis, the last integral tends to

$$\int_{1/k}^{\infty} [(z^2 - 1)(k^2 z^2 - 1)]^{-1/2} dz = K.$$

Hence the image of the segment of the real axis from $1/k$ to $+\infty$ is the straight line segment from $K + iK'$ to iK'. By considering the case $\zeta < 0$, we see that the image of the real axis under $\zeta \mapsto F(\zeta; k)$, where the path of integration is along the real axis and along small semi-circle in the upper-half-plane around the points $z = \pm 1$ and $\pm 1/k$, is the rectangle with vertices $\pm K$ and $iK' \pm K$. Furthermore, the upper half-plane is mapped in a one-to-one fashion to the interior of this rectangle.

By repeating the above argument but using a path of integration that follows small semi-circles in the lower half-plane about the singularities, we see that the lower half-plane is mapped to the rectangle with vertices $\pm K$ and $-iK' \pm K$. Furthermore, by looping around each of the singularities a number of times, it is possible to map the upper or lower half-planes to different rectangles. In this way we can cover the entire complex plane with rectangles each of which is the image of either the upper or lower half-planes. This means that we can define a single-valued inverse function $\mathrm{sn}(z; k)$.

The function $\mathrm{sn}(z; k)$ is an elliptic function with periods $4K$ and $2iK'$. Note that

$$\left(\frac{d\,\mathrm{sn}(z; k)}{dz} \right)^2 = (1 - \mathrm{sn}^2(z; k))(1 - k^2 \mathrm{sn}^2(z; k)).$$

2.5. Jacobi elliptic functions

The Jacobi elliptic function $\mathrm{sn}(z; k)$ is a generalization of $\sin z$. In fact, $\mathrm{sn}(z; 0) = \sin(z)$. There are two other important Jacobi elliptic functions, $\mathrm{cn}(z; k)$, which generalizes $\cos z$, and $\mathrm{dn}(z; k)$, which does not have a trigonometric analogue. (The natural generalization of $\tan z$ is $\mathrm{sc}(z;k) :=$ $\mathrm{sn}(z;k)/\mathrm{cn}(z;k)$.) These Jacobi elliptic functions can be defined in terms of sn by

$$\mathrm{sn}^2(z;k) + \mathrm{cn}^2(z;k) = 1, \quad \mathrm{cn}(0;k) = 1$$

and

$$k^2\mathrm{sn}^2(z;k) + \mathrm{dn}^2(z;k) = 1, \quad \mathrm{dn}(0;k) = 1.$$

2.6.   Exercises

(1) Let f be an elliptic function such that $f(z) \neq 1$ for all $z \in \mathbb{C}$. Show that f is a constant.

(2) Find the general solution of the equation

$$\frac{d^2y}{dz^2} = 6y^2 + \kappa,$$

where κ is a constant, in terms of the Weierstrass $\wp$ function.

(3) Repeat the conformal mapping argument in Section 6.6 in the case in which k is real but $k > 1$. In particular, what is the image of the upper half-plane in this case?

(4) Solve the initial value problem

$$\frac{d^2y}{dz^2} = 3y^3 - 9y, \quad y(0) = 0, \quad y'(0) = 2.$$

(5) Let

$$u(z) = e_3 + \frac{e_1 - e_3}{\mathrm{sn}^2(\lambda z; k)},$$

where

$$\lambda^2 = e_1 - e_3, \quad k^2 = \frac{e_2 - e_3}{e_1 - e_3}, \quad e_1 + e_2 + e_3 = 0,$$

and the e_j's are all distinct. Show that

$$(u')^2 = 4(u - e_1)(u - e_2)(u - e_3).$$

Hence show that $u(z) = \wp(z; g_2, g_3)$ for appropriate constants g_2 and g_3.

(6) Use an appropriate Möbius transformation to find the general solution of

$$\left(\frac{dy}{dz}\right)^2 = 4(y+1)y(y-2)(y-3)$$

in terms of Jacobi elliptic functions.

3. Linear Differential Equations

The point $z = z_0$ is called an ordinary point of the linear ODE

$$\frac{d^n y}{dz^n} + p_1(z)\frac{d^{n-1} y}{dz^{n-1}} + \cdots + p_n(z)y = 0 \tag{3.1}$$

if the coefficient functions p_k, $k = 1, \ldots, n$, are analytic in a neighbourhood of $z = z_0$. Otherwise $z = z_0$ is called a singular point of the equation.

A solution of equation (3.1) can be analytically continued along any curve that avoids the singular points of the equation. So the singularities of a solution y of equation (3.1) can only occur at its singular points. This is in stark contrast to the case of nonlinear equations.

3.1. *First-order linear equations*

Consider the first-order linear homogeneous ODE

$$\frac{dy}{dz} + Q(z)y = 0, \tag{3.2}$$

where Q is analytic in some neighbourhood of the point $z = a$. The general solution of equation (3.2) is

$$y(z) = \kappa \exp\left(-\int_\gamma Q(\zeta)d\zeta\right), \tag{3.3}$$

where $\kappa = y(a)$ and in general the value of $y(z)$ depends on the path of integration γ from a to z. If Q is everywhere analytic then y will have no singularities. Suppose that Q has a pole of order q at $z = z_0$. On substituting the expansion

$$Q(z) = \sum_{n=-q}^{\infty} Q_n(z - z_0)^n, \quad Q_{-q} \neq 0,$$

into equation (3.3) we see that, up to a multiplicative constant, the solution has the form

$$y(z) = f(z)g(z)h(z),$$

where

$$f(z) = \exp\left(-\sum_{m=0}^{\infty} \frac{Q_m}{m+1}(z - z_0)^{m+1}\right), \quad g(z) = (z - z_0)^r,$$

$$h(z) = \exp\left(-\sum_{m=-q}^{-2} \frac{Q_m}{m+1}(z - z_0)^{m+1}\right)$$

and $r = -Q_{-1}$. Clearly f is analytic at $z = z_0$, g is either meromorphic or branched at $z = z_0$ and, provided that $q \geq 2$, h has an essential singularity at $z = z_0$.

In the above analysis we were able to solve equation (3.2) explicitly. In general, this will not be the case when we consider higher-order equations. For this reason we would like to construct some kind of useful series representation of the solution y near a pole of Q without using the explicit solution. The above argument shows that if $q = 1$ then $y(z)$ has a series expansion about $z = z_0$ of the form

$$y(z) = \sum_{n=0}^{\infty} a_n (z - z_0)^{n+r}, \quad a_0 \neq 0. \tag{3.4}$$

If $q > 1$ then $(z - z_0)^{-r} y(z)$ has an essential singularity at $z = z_0$ and the generation of local series expansions directly from the equation itself becomes much more complicated. If $q = 1$ then $z = z_0$ is said to be a regular singular point of equation (3.2).

3.2. Regular singular points

Equation (3.1) is said to have a *regular singular point* (or a *Fuchsian singularity*) at $z = z_0$ if for all $j = 1, \ldots, n$, p_j is meromorphic in a neighbourhood of the singular point $z = z_0$ and if p_j has a pole at $z = z_0$ it has order no greater than j. In particular, a linear homogeneous second-order ODE for $y(z)$ has a regular singularity at $z = z_0$ if it has the form

$$\frac{d^2 y}{dz^2} + \sum_{n=0}^{\infty} p_n (z - z_0)^{n-1} \frac{dy}{dz} + \sum_{n=0}^{\infty} q_n (z - z_0)^{n-2} y = 0, \tag{3.5}$$

where at least one of the constants p_0, q_0, q_1 is not zero. Equation (3.5) has a solution of the form (3.4), where r is a root of the *indicial equation*

$$r(r - 1) + p_0 r + q_0 = 0. \tag{3.6}$$

Moreover if the roots of equation (3.6) do not differ by an integer then there are two independent solutions of the form (3.4), corresponding to the two values of r.

For an equation of the form (3.1), the point at infinity is called an ordinary point, a regular singular point, or an irregular singular point if the resulting linear equation for $w(x) = y(z)$, $x = 1/z$ has an ordinary point, a regular singular point, or an irregular singular at $x = 0$ respectively. In particular, under this change of variables, the equation

$$\frac{d^2y}{dz^2} + P(z)\frac{dy}{dz} + Q(z)y = 0, \tag{3.7}$$

becomes

$$\frac{d^2w}{dx^2} + \left(\frac{2}{x} - \frac{P(1/x)}{x^2}\right)\frac{dw}{dx} + \frac{Q(1/x)}{x^4}w = 0.$$

It follows that equation (3.7) has an ordinary point at infinity if and only if

$$P(z) = \frac{2}{z} + O\left(\frac{1}{z^2}\right), \qquad Q(z) = O\left(\frac{1}{z^4}\right), \quad z \to \infty. \tag{3.8}$$

If equation (3.7) has a singularity at infinity then this singularity is regular if and only if

$$P(z) = O\left(\frac{1}{z}\right), \qquad Q(z) = O\left(\frac{1}{z^2}\right), \quad z \to \infty.$$

3.3.   *The hypergeometric equation*

The hypergeometric equation is

$$z(1-z)\frac{d^2y}{dz^2} + [c - (a+b+1)z]\frac{dy}{dz} - ab\,y = 0, \tag{3.9}$$

where a, b, and c are constants. It is one of the most important equations in mathematics and arises in many applications.

The hypergeometric equation is an example of a Fuchsian equation. An ODE is said to be Fuchsian if all its singular points in the extended complex plane are regular (i.e. Fuchsian). Many of the great early mathematical analysts including Gauss, Riemann, Fuchs, Kummer, and Klein produced detailed studies of Fuchsian equations. The hypergeometric equation (3.9) has singular points at $z = 0$, $z = 1$, and $z = \infty$. Fuchsian equations with fewer than three singular points can be solved explicitly (see the exercises). We will see that any Fuchsian equation with exactly three singular points can be mapped to equation (3.9) by a simple change of variables. Fuchsian equations with four or more singular points are much more difficult to analyse than those with only three.

We will consider the global consequences of the branching of solutions of the hypergeometric equation. In particular, we will see how a solution transforms as it is analytically continued around any closed curve in the plane which avoids the singularities at $z = 0$, $z = 1$, and $z = \infty$.

3.4.   Riemann's equation

In this section we will derive the general form of a Fuchsian equation with three (distinct) regular singular points at x_1, x_2, and x_3 (initially we will assume that x_1, x_2, and x_3 are finite). Such an equation necessarily has the form

$$\frac{d^2u}{dx^2} + \frac{p(x)}{(x-x_1)(x-x_2)(x-x_3)}\frac{du}{dx} + \frac{q(x)}{[(x-x_1)(x-x_2)(x-x_3)]^2}u = 0,$$

where p and q are polynomials. Since $x = \infty$ is an ordinary point it follows from condition (3.8) that for $x \to \infty$, the coefficient of du/dx must be $2x^{-1} + O(x^{-2})$ and the coefficient of u must be $O(x^{-4})$. Hence

$$\frac{d^2u}{dx^2} + \left(\frac{A_1}{x-x_1} + \frac{A_2}{x-x_2} + \frac{A_3}{x-x_3}\right)\frac{du}{dx}$$
$$+ \left(\frac{B_1}{x-x_1} + \frac{B_2}{x-x_2} + \frac{B_3}{x-x_3}\right)\frac{u}{(x-x_1)(x-x_2)(x-x_3)} = 0, \quad (3.10)$$

where $A_1 + A_2 + A_3 = 2$. The indicial equation for (3.10) at the point $x = x_i$ is

$$r^2 + (A_i - 1)r + \frac{B_i}{\prod_{j \neq i}(x_i - x_j)} = 0.$$

Hence if r_i and s_i are the roots of the indicial equation at x_i, then $r_i + s_i = 1 - A_i$ and $r_i s_i = B_i/\prod_{j \neq i}(x_i - x_j)$. Hence equation (3.10) becomes the Riemann equation

$$\frac{d^2u}{dx^2} + \left\{\frac{1-r_1-s_1}{x-x_1} + \frac{1-r_2-s_2}{x-x_2} + \frac{1-r_3-s_3}{x-x_3}\right\}\frac{du}{dx}$$
$$+ \left\{\frac{r_1s_1(x_1-x_2)(x_1-x_3)}{x-x_1} + \frac{r_2s_2(x_2-x_3)(x_2-x_1)}{x-x_2}\right.$$
$$\left.+ \frac{r_3s_3(x_3-x_1)(x_3-x_2)}{x-x_3}\right\}\frac{u}{(x-x_1)(x-x_2)(x-x_3)} = 0, \quad (3.11)$$

where $r_1 + r_2 + r_3 + s_1 + s_2 + s_3 = 1$. We derived equation (3.11) under the assumption that x_1, x_2, and x_3 are finite. However, if we take the limit $x_3 \to \infty$ in equation (3.11) then we obtain the general form of a Fuchsian equation with regular singular points at x_1, x_2, and ∞. In particular, if we choose $x_1 = 0$, $x_2 = 1$, $x_3 = \infty$, $r_1 = 0$ and $r_2 = 0$, then equation (3.11) becomes the hypergeometric equation (3.9), where we have set $s_1 = 1 - c$, $s_2 = c - a - b$, $r_3 = a$, $s_3 = b$, and $x = z$.

The general solution of Riemann's equation (3.11) is denoted by

$$
P \left\{
\begin{array}{cccc}
x_1 & x_2 & x_3 & \\
r_1 & r_2 & r_3 & x \\
s_1 & s_2 & s_3 &
\end{array}
\right\},
$$

where r_i and s_i are the roots of the indicial equation at the regular singular point x_i. So the general solution of the hypergeometric equation (3.9) is denoted by

$$
P \left\{
\begin{array}{cccc}
0 & 1 & \infty & \\
0 & 0 & a & z \\
1-c & c-a-b & b &
\end{array}
\right\}.
\tag{3.12}
$$

We will now show that any Fuchsian equation with three singular points (i.e. Riemann's equation) can be mapped to the hypergeometric equation (3.9).

The Möbius transformation

$$
x \mapsto z = \frac{(x_2 - x_3)(x - x_1)}{(x_2 - x_1)(x - x_3)}
\tag{3.13}
$$

maps the regular singular points x_1, x_2 and x_3 to 0, 1 and ∞ respectively while maintaining the Fuchsian nature of the equation. Let $y(z) = u(x)$ be the general solution of the transformed equation. Note that

$$
z := \begin{cases}
\dfrac{x_2 - x_3}{(x_3 - x_1)(x_1 - x_2)}(x - x_1) + O\left((x - x_1)^2\right); & x \to x_1, \\[3mm]
1 + \dfrac{x_3 - x_1}{(x_1 - x_2)(x_2 - x_3)}(x - x_2) + O\left((x - x_1)^2\right); & x \to x_2, \\[3mm]
-\dfrac{(x_2 - x_3)(x_3 - x_1)}{x_1 - x_2}(x - x_3)^{-1} + O(1) & x \to \infty.
\end{cases}
$$

It follows that the roots of the indicial equations for equation (3.11) at the singular points $x = x_1$, $x = x_2$ and $x = x_3$ are the same as those for the transformed equation at $z = 0$, $z = 1$, and $z = \infty$ respectively. Furthermore, one of the roots of the indicial equation at $z = 0$ can be normalised to 0 by dividing $y(z)$ by z^{r_1}. Similarly one of the roots of the indicial equation at $z = 1$ can be normalised to 0 by dividing $y(z)$ by $(z-1)^{r_2}$. This shows that the general solution of Riemann's equation (3.11) is

$$P\left\{\begin{matrix} x_1 & x_2 & x_3 \\ r_1 & r_2 & r_3 & x \\ s_1 & s_2 & s_3 \end{matrix}\right\} = z^{r_1}(z-1)^{r_2}P\left\{\begin{matrix} 0 & 1 & \infty \\ 0 & 0 & r_1+r_2+r_3 & z \\ s_1-r_1 & s_2-r_2 & r_1+r_2+s_3 \end{matrix}\right\},$$

which can be written in terms of the solution of the hypergeometric equation as

$$\left(\frac{x-x_1}{x-x_3}\right)^{r_1}\left(\frac{x-x_2}{x-x_3}\right)^{r_2}P\left\{\begin{matrix} 0 & 1 & \infty \\ 0 & 0 & a & \dfrac{(x_2-x_3)(x-x_1)}{(x_2-x_1)(x-x_3)} \\ 1-c & c-a-b & b \end{matrix}\right\},$$

$$(3.14)$$

where $a = r_1 + r_2 + r_3$, $b = r_1 + r_2 + s_3$, and $c = 1 + r_1 - s_1$.

3.5. Series solutions

In this section we will find series expansions of solutions of the hypergeometric equation (3.9) about its singular points. We begin by finding series expansions for solutions about $z = 0$. On substituting $y(z) = \sum_{n=0}^{\infty} \alpha_n z^{n+r}$ into equation (3.9) we obtain

$$0 = \sum_{n=0}^{\infty} \alpha_n[(n+r)(n+r-1)z^{n+r-1} - (n+r)(n+r-1)z^{n+r}$$
$$+ c(n+r)z^{n+r-1} - (a+b+1)(n+r)z^{n+r} - abz^{n+r}]$$
$$= \alpha_0 r(r+c-1)z^{r-1} + \sum_{n=0}^{\infty}[\alpha_{n+1}(n+r+1)(n+r+c)$$
$$- \alpha_n(n+r+a)(n+r+b)]z^{n+r}.$$

Equating the coefficient of z^{r-1} to zero gives the indicial equation, which has roots $r = 0 =: r_{01}$ and $r = 1-c =: r_{02}$. We will assume that the roots of the indicial equation at each singularity do not differ by integers. Equating the coefficient of z^{r+n} to zero gives the recurrence relation

$$\alpha_{n+1} = \frac{(n+r+a)(n+r+b)}{(n+r+1)(n+r+c)}\alpha_n \quad n = 0, 1, \dots. \tag{3.15}$$

For the case $r = r_{01} = 0$ this becomes

$$\alpha_{n+1} = \frac{(n+a)(n+b)}{(n+1)(n+c)}\alpha_n \quad n = 0, 1, \ldots,$$

yielding

$$\alpha_1 = \frac{ab}{1 \cdot c}\alpha_0, \quad \alpha_2 = \frac{(a+1)(b+1)}{2 \cdot (c+1)}\alpha_1 = \frac{a(a+1)b(b+1)}{1 \cdot 2 \cdot c(c+1)}\alpha_0,$$

$$\alpha_n = \frac{a(a+1)\cdots(a+n-1)\cdot b(b+1)\cdots(b+n-1)}{n!c(c+1)\cdots(c+n-1)}\alpha_0. \tag{3.16}$$

Substituting $r = r_{02} = 1 - c$ in the recurrence relation (3.15) gives

$$\hat{\alpha}_{n+1} = \frac{(n+a-c+1)(n+b-c+1)}{(n+1)(n+2-c)}\hat{\alpha}_n, \tag{3.17}$$

which can be obtained by replacing a with $a-c+1$, b with $b-c+1$, and c with $2-c$, in the recurrence relation (3.15). So two independent solutions of the hypergeometric equation (3.9) are

$$y_{01}(z) = \sum_{n=0}^{\infty} \alpha_n z^n = F(a, b, c; z),$$

and

$$y_{02}(z) = z^{1-c}\sum_{n=0}^{\infty} \hat{\alpha}_n z^n = z^{1-c}F(a-c+1, b-c+1, 2-c; z),$$

where

$$F(a, b, c; z) := 1 + \sum_{n=1}^{\infty} \frac{(a)_n(b)_n}{n!(c)_n}z^n$$

and $(c)_n = \Gamma(c+n)/\Gamma(c)$ (i.e. $(c)_0 = 1$ and $(c)_n = c(c+1)\cdots(c+n-1)$, $n = 1, 2, \ldots$). Hence the hypergeometric series F can also be written as

$$F(a, b, c; z) = \frac{\Gamma(c)}{\Gamma(a)\Gamma(b)}\sum_{n=0}^{\infty} \frac{\Gamma(a+n)\Gamma(b+n)}{n!\Gamma(c+n)}z^n.$$

From equation (3.16) we have

$$\left|\frac{\alpha_{n+1}z^{n+1}}{\alpha_n z^n}\right| = \left|\frac{(n+a)(n+b)}{(n+1)(n+c)}\right||z| \to |z|$$

as $n \to \infty$. So the series converges absolutely for $|z| < 1$ by the Ratio Test.

Series expansions about the other singularities show that two independent solutions near $z = 1$ are

$$y_{11} = F(a, b, a + b - c + 1; 1 - z), \tag{3.18}$$

$$y_{12} = (1 - z)^{c-a-b} F(c - a, c - b, c - a - b + 1; 1 - z) \tag{3.19}$$

and two independent solutions near $z = \infty$ are

$$y_{\infty 1} = z^{-a} F(a, a - c + 1, a - b + 1; z^{-1}), \tag{3.20}$$

$$y_{\infty 2} = z^{-b} F(b, b - c + 1, b - a + 1; z^{-1}). \tag{3.21}$$

3.6. Kummer's solutions

In Section 3.4 we saw that any second-order Fuchsian equation with three
regular singular points (3.11) can be transformed to the hypergeometric
equation (3.9) by first using a Möbius transformation to map the three
singular points to 0, 1 and ∞ and then replacing u by $z^{r_1}(z - 1)^{r_2} y(z)$
where r_1 and r_2 are roots of the indicial equations at $z = 0$ and $z = 1$
respectively.

There are exactly six Möbius transformations that map the set $\{0, 1, \infty\}$
onto itself. These transformations form a group called the anharmonic group
and are given in the table below. So starting from the hypergeometric equation (3.9) with fixed values of a, b, and c, we can write down six equations
of the form (3.11) that arise from the six permutations of $(0, 1, \infty)$ given by
the anharmonic transformations. For each of these six equations there are
(generically) two roots r_{01}, r_{02} of the indicial equation at $z = 0$ and two
roots r_{11}, r_{12} of the indicial equation at $z = 1$, giving us four factors of the
form $y(z) = z^{-r_{0m}}(z - 1)^{-r_{1n}} u(z)$, $m, n = 1, 2$, such that y again satisfies
a hypergeometric equation (3.9), however, the parameters a, b, and c will
have changed in general. This gives us a total of 24 solutions to the hypergeometric equation that can be expressed in terms of any given solution,
say $y_{01}(z) = F(a, b, c; z)$.

Transformation $\zeta(z)$	z	$1 - z$	$\frac{1}{z}$	$\frac{1}{1-z}$	$\frac{z}{z-1}$	$\frac{z-1}{z}$
$\zeta(0)$	0	1	∞	1	0	∞
$\zeta(1)$	1	0	1	∞	∞	0
$\zeta(\infty)$	∞	∞	0	0	1	1

For the hypergeometric equation there are six ways of labelling the
points 0, 1, ∞ as z_1, z_2, z_3, corresponding to the elements of the anharmonic

group. There are two ways of labelling the roots of the indicial equation at $z = 0$ (i.e. either $r_j = 0$ and $s_j = 1 - c$ or $s_j = 0$ and $r_j = 1 - c$). Similarly there are two ways of labelling the roots of the indicial equation at $z = 1$ (0 and $c - a - b$). In this way we generate 24 solutions of the hypergeometric equation. We have

$$P \left\{ \begin{matrix} z_1 & z_2 & z_3 \\ r_1 & r_2 & r_3 & z \\ s_1 & s_2 & s_3 \end{matrix} \right\} = \left(\frac{z - z_1}{z - z_3} \right)^{r_1} \left(\frac{z - z_2}{z - z_3} \right)^{r_2}$$

$$\times P \left\{ \begin{matrix} 0 & 1 & \infty \\ 0 & 0 & \hat{a} & \frac{(z_2 - z_3)(z - z_1)}{(z_2 - z_1)(z - z_3)} \\ 1 - \hat{c} & \hat{c} - \hat{a} - \hat{b} & \hat{b} \end{matrix} \right\},$$

where $\hat{a} = r_1 + r_2 + r_3$, $\hat{b} = r_1 + r_2 + s_3$, and $\hat{c} = 1 + r_1 - s_1$. Note that this is a relationship between general solutions so we ignore overall multiplicative constants. When we take a limit, e.g. $z_3 \to \infty$, we can rescale the right side with a factor of $z_3^{r_1 + r_2}$.

Let us follow the above procedure to find three other expressions for $y_{01} = F(a, b, c; z)$. This solution is characterised by the fact that it is analytic at $z = 0$, where it has the value 1. If we choose the Möbius transformation to be the identity (i.e. $z_1 = 0$, $z_2 = 1$, $z_3 = \infty$), then in order to get a solution analytic at $z = 0$ we should choose $r_1 = 0$. The choice $r_2 = 0$ does not give any new solution. The choice $r_2 = c - a - b$ gives the solution $(z - 1)^{c-a-b} F(c - b, c - a, c; z)$. This solution is analytic at $z = 0$. Multiplication by $(-1)^{c-a-b}$ gives a solution which is 1 at $z = 0$ (for an appropriate choice of branch).

Apart from the identity, the only anharmonic transformation that maps $z = 0$ to itself is $z \mapsto \frac{z}{z-1}$. On making this transformation in the hypergeometric equation and multiplying by one of two powers of $(1 - z)$, we again obtain a solution of the hypergeometric equation (with different parameters) which is 1 at $z = 0$. So we have four different expressions for y_{01}. Notice that this second pair of solutions converge in the disc $|z| < |z - 1|$, providing a semi-explicit analytic continuation. Using similar reasoning for the other solutions, we find a total of 24 solutions which are naturally grouped into six sets of identical functions.

$$y_{01} = F(a, b, c; z)$$

$$= (1 - z)^{c-a-b} F(c - a, c - b, c; z)$$

$$= (1-z)^{-a} F\left(a, c-b, c; \frac{z}{z-1}\right)$$

$$= (1-z)^{-b} F\left(b, c-a, c; \frac{z}{z-1}\right),$$

$$y_{02} = z^{1-c} F(a-c+1, b-c+1, 2-c; z)$$

$$= z^{1-c}(1-z)^{c-a-b} F(1-a, 1-b, 2-c; z)$$

$$= z^{1-c}(1-z)^{c-a-1} F\left(a-c+1, 1-b, 2-c; \frac{z}{z-1}\right)$$

$$= z^{1-c}(1-z)^{c-b-1} F\left(b-c+1, 1-a, 2-c; \frac{z}{z-1}\right),$$

$$y_{11} = F(a, b, a+b-c+1; 1-z)$$

$$= z^{1-c} F(a-c+1, b-c+1, a+b-c+1; 1-z)$$

$$= z^{-a} F\left(a, a-c+1, a+b-c+1; \frac{z-1}{z}\right)$$

$$= z^{-b} F\left(b, b-c+1, a+b-c+1; \frac{z-1}{z}\right),$$

$$y_{12} = (1-z)^{c-a-b} F(c-a, c-b, c-a-b+1; 1-z)$$

$$= (1-z)^{c-a-b} z^{1-c} F(1-a, 1-b, c-a-b+1; 1-z)$$

$$= (1-z)^{c-a-b} z^{a-c} F\left(1-a, c-a, c-a-b+1; \frac{z-1}{z}\right)$$

$$= (1-z)^{c-a-b} z^{b-c} F\left(1-b, c-b, c-a-b+1; \frac{z-1}{z}\right),$$

$$y_{\infty 1} = z^{-a} F(a, a-c+1, a-b+1, z^{-1})$$

$$= z^{-a} \left(\frac{1-z}{z}\right)^{c-a-b} F(1-b, c-b, a-b+1, z^{-1})$$

$$= z^{-a} \left(\frac{1-z}{z}\right)^{-a} F\left(a, c-b, a-b+1; \frac{1}{1-z}\right)$$

$$= z^{-a} \left(\frac{1-z}{z}\right)^{c-a-1} F\left(a-c+1, 1-b, a-b+1; \frac{1}{1-z}\right),$$

$$y_{\infty 2} = z^{-b} F(b, b - c + 1, b - a + 1, z^{-1})$$

$$= z^{-b} \left(\frac{1-z}{z} \right)^{c-a-b} F(1 - a, c - a, b - a + 1, z^{-1})$$

$$= z^{-b} \left(\frac{1-z}{z} \right)^{-a} F\left(b, c - a, b - a + 1; \frac{1}{1-z} \right)$$

$$= z^{-b} \left(\frac{1-z}{z} \right)^{c-a-1} F\left(b - c + 1, 1 - a, b - a + 1; \frac{1}{1-z} \right).$$

3.7. *Integral representations*

The hypergeometric equation (3.9) can be written as $Ly = 0$, where

$$L = z(1 - z)\frac{\partial^2}{\partial z^2} + [c - (a + b + 1)z]\frac{\partial}{\partial z} - ab.$$

From the identity

$$L\left[t^{b-1}(1 - t)^{c-b-1}(1 - tz)^{-a} \right] = -a\frac{\partial}{\partial t}\left[t^b (1 - t)^{c-b}(1 - tz)^{-a-1} \right],$$

it follows that

$$\int_\gamma t^{b-1}(1 - t)^{c-b-1}(1 - tz)^{-a}\, dt \tag{3.22}$$

solves equation (3.9) provided the integral converges and the path of integration γ in equation (3.22) is chosen such that either γ is closed on the Riemann surface of the integrand or terminates at points where $t^b(1 - t)^{c-b}(1 - tz)^{-a-1}$ vanishes.

In particular, if $\Re(c) > \Re(b) > 0$, then

$$y(z) := \int_0^1 t^{b-1}(1 - t)^{c-b-1}(1 - tz)^{-a}\, dt \tag{3.23}$$

is a solution of the hypergeometric equation. Note that y is analytic in a neighbourhood of $z = 0$, so $y(z) = \kappa F(a, b, c; z)$, for some constant κ. Setting $z = 0$ and noting that $F(a, b, c; 0) = 1$ gives

$$\kappa = y(0) = \int_0^1 t^{b-1}(1 - t)^{c-b-1}\, dt = \frac{\Gamma(b)\Gamma(c - b)}{\Gamma(c)}.$$

Hence

$$F(a, b, c; z) = \frac{\Gamma(c)}{\Gamma(b)\Gamma(c - b)} \int_0^1 t^{b-1}(1 - t)^{c-b-1}(1 - tz)^{-a} \, dt. \qquad (3.24)$$

It follows that

$$F(a, b, c; 1) = \frac{\Gamma(c)\Gamma(c - a - b)}{\Gamma(c - a)\Gamma(c - b)}. \qquad (3.25)$$

3.8. *Monodromy of the hypergeometric equation*

In this section we will consider the result of analytically continuing a solution y of the hypergeometric equation around any closed curve in the complex plane that does not pass through 0, 1, or ∞. In general, if we start at some point z_0 on γ and analytically continue the solution along γ back to z_0, we will not return to the same value due to branching at $z = 0$, $z = 1$, and $z = \infty$. Note that we are free to deform γ so long as it does not cross these singular points. Hence, to analyse the analytic continuation of y along any closed curve, we need only consider its analytic continuation around $z = 0$ and $z = 1$. Analytic continuation around $z = \infty$ is equivalent to analytic continuation around $z = 0$ and $z = 1$.

We will denote by γ_0 and γ_1 the circles of radius ϵ (where $0 < \epsilon < 1$) traversed in the positive (anti-clockwise) direction, with centres $z = 0$ and $z = 1$ respectively. We begin by considering analytic continuation around $z = 0$. For this purpose, it is convenient to use the basis of solutions

$$y_{01}(z) = F(a, b, c; z),$$
$$y_{02}(z) = z^{1-c}F(a - c + 1, b - c + 1, 2 - c; z).$$

Recall that F is analytic in a neighbourhood of $z = 0$, so y_{01} is unchanged on analytic continuation around γ_0. However, on analytic continuation around γ_0, the original value of the factor z^{1-c} is multiplied $e^{-2\pi i c}$. We summarise this by writing

$$\gamma_0 : \quad (y_{01} \ y_{02}) \mapsto (y_{01} \ e^{-2\pi i c} y_{02}).$$

Similarly, near $z = 1$ it is convenient to use the basis of solutions

$$y_{11}(z) = F(a, b, a + b - c + 1; 1 - z),$$
$$y_{12}(z) = (1 - z)^{c-a-b}F(c - a, c - b, c - a - b + 1; 1 - z).$$

The circle γ_1 can be parameterised as $z = 1 + \epsilon e^{i\theta}$, where θ ranges from 0 to 2π. The only branching in the continuation of y_{11} and y_{12} around γ_1 comes from the factor $(1 - z)^{c-a-b}$. This leads to

$$\gamma_1 : \quad \left(y_{11} \ y_{12}\right) \mapsto \left(y_{11} \ e^{2\pi i(c-a-b)} y_{12}\right).$$

In order to track the behaviour of a particular solution around a curve looping around both $z = 0$ and $z = 1$, we need to choose a single pair of basis solutions and relate all others to this pair. A suitable pair is y_{01} and y_{11}, which are chosen because they are both branched around only one of the two singular points (it remains to be seen that these two solutions are independent). Since y_{11} and y_{12} form a basis, there exist constants α and β such that

$$y_{01} = \alpha y_{11}(z) + \beta y_{12}(z). \tag{3.26}$$

This is equivalent to

$$F(a, b, c; z) = \alpha F(a, b, a + b - c + 1; 1 - z)$$
$$+ \beta(1 - z)^{c-a-b} F(c - a, c - b, c - a - b + 1; 1 - z). \tag{3.27}$$

In the case $\Re(c - a - b) > 0$, we substitute $z = 1$ in equation (3.27) to give

$$F(a, b, c; 1) = \alpha F(a, b, a + b - c + 1; 0).$$

Using $F(a, b, c; 0) = 1$ and equation (3.25), we find

$$\alpha = \frac{\Gamma(c)\Gamma(c - a - b)}{\Gamma(c - a)\Gamma(c - b)}.$$

From equation (3.26), we see that after y_{01} is continued along γ_1, we have

$$\gamma_1 : \quad y_{01} \mapsto \alpha y_{11} + \beta e^{2\pi i(c-a-b)} y_{12} = e^{2\pi i(c-a-b)} y_{01} + \alpha(1 - e^{2\pi i(c-a-b)}) y_{11}.$$

Similarly, it can be shown that

$$y_{11} = \gamma y_{01} + \delta y_{02}, \tag{3.28}$$

where

$$\gamma = \frac{\Gamma(a + b - c + 1)\Gamma(1 - c)}{\Gamma(b - c + 1)\Gamma(a - c + 1)}. \tag{3.29}$$

Hence, after analytic continuation along γ_0, we have

$$\gamma_0: \quad y_{11} \mapsto \gamma(1 - e^{-2\pi i c})y_{01} + e^{-2\pi i c}y_{11}. \tag{3.30}$$

Now using the identity $\Gamma(1 - z)\Gamma(z) = \pi/\sin \pi z$, we have

$$\alpha\gamma = \frac{\sin \pi(c - a) \sin \pi(c - b)}{\sin \pi c \sin \pi(c - a - b)} = \frac{(1 - e^{2i\pi(c-a)})(1 - e^{2i\pi(b-c)})}{e^{2i\pi b}(1 - e^{2i\pi(c-a-b)})(1 - e^{-2i\pi c})}.$$

So finally, if we choose

$$y_A(z) := y_{01}(z) \quad \text{and} \quad y_B = \alpha \frac{1 - e^{2i\pi(c-a-b)}}{1 - e^{2i\pi(c-a)}} y_{11}(z)$$

and let $\mathbf{Y} := (y_A, y_B)$, then

$$\gamma_0: \quad \mathbf{Y} \mapsto \mathbf{Y}\mathbf{M}_0, \qquad \gamma_1: \quad \mathbf{Y} \mapsto \mathbf{Y}\mathbf{M}_1,$$

where

$$\mathbf{M}_0 := \begin{pmatrix} 1 & e^{-2\pi i b}(1 - e^{2i\pi(b-c)}) \\ 0 & e^{-2\pi i c} \end{pmatrix}$$

and

$$\mathbf{M}_1 := \begin{pmatrix} e^{2\pi i(c-a-b)} & 0 \\ 1 - e^{2\pi i(c-a)} & 1 \end{pmatrix}$$

are the monodromy matrices around $z = 0$ and $z = 1$ respectively.

Note that every closed loop that avoids $z = 0$, $z = 1$, and $z = \infty$ can be divided into successive loops γ_0 and γ_1, which may be traversed in the positive or negative directions. After analytic continuation along such a curve, $\mathbf{Y}$ returns to the value

$$\mathbf{Y} \mapsto \mathbf{Y}\mathbf{M}, \quad \mathbf{M} = \mathbf{M}_0^{p_1}\mathbf{M}_1^{q_1}\mathbf{M}_0^{p_2}\mathbf{M}_1^{q_2} \cdots \mathbf{M}_0^{p_n}\mathbf{M}_1^{q_n},$$

for some integers $p_1, \ldots, p_n$ and $q_1, \ldots, q_n$.

3.9. Exercises

(1) Show that

$$\frac{d}{dz}F(a, b, c; z) = \frac{ab}{c}F(a + 1, b + 1, c + 1; z),$$

and

$$\frac{d^n}{dz^n}F(a,b,c;z) = \frac{(a)_n(b)_n}{(c)_n}F(a+n,b+n,c+n;z).$$

(2) Show that $F(1,1/2,3/2;-z^2) = z^{-1}\tan^{-1}z$.

(3) Express the general solution of the equation

$$x(x-2)\frac{d^2u}{dx^2} + \left(\frac{4}{3}x - 1\right)\frac{du}{dx} + \frac{1}{72}u = 0$$

in terms of F.

(4) (a) Show that the only second-order linear homogeneous ODE with rational coefficients that has regular singularities at $z=a$ and $z=b$ ($a \neq b$) and no singularity at $z=\infty$ (i.e. $z=\infty$ is an ordinary point) is

$$\frac{d^2y}{dz^2} + \frac{2z+\mu}{(z-a)(z-b)}\frac{dy}{dz} + \frac{\nu}{(z-a)^2(z-b)^2}y = 0,$$

where μ and ν are constants.

(b) Solve the equation

$$\frac{d^2y}{dz^2} + \frac{2z+2}{z(z-1)}\frac{dy}{dz} + \frac{2}{z^2(z-1)^2}y = 0$$

by using a fractional linear (i.e. Möbius) transformation to map the regular singular points at $z = 0$ and $z = 1$ to 0 and ∞ respectively and solving the resulting Euler equation.

4. Matrix Fuchsian Equations

A linear equation

$$\frac{d\Psi}{dz} = A(z)\Psi, \tag{4.1}$$

where A is an $m \times m$ matrix, is said to have a Fuchsian singularity at $z = z_0$ if $A(z)$ has a simple pole at $z = z_0$. Hilbert's 21st problem is the following. Given a finite collection of distinct points $z_1, \ldots, z_n$ in the extended complex plane, and suitable matrices $M_1, \ldots, M_n$, is there a Fuchsian equation of the form (4.1) where the z_js are the singular points and the M_j are the corresponding monodromy matrices? It was believed that this problem had been solved in the affirmative by Plemelj a few years after the problem was first posed. However, a counterexample was discovered by Bolibruch (see [2]). Plemelj had actually constructed a regular, rather than Fuchsian, equation.

Let $A(z)$ have a simple pole at $z = z_0$. Suppose that in a neighbourhood of $z = z_0$, $A(z) = \frac{A_0}{z - z_0} + O(1)$, where no two eigenvalues of A_0 differ by an integer. Then it can be shown that the differential equation (4.1) has a fundamental matrix solution of the form

$$\Psi(z) = \Phi(z)z^{A_0}, \quad \Phi(z_0) = I,$$

where Φ is analytic at $z = z_0$. Note that $z^{A_0} = \exp(A_0 \ln z)$.

4.1. *Isomonodromic deformations*

The following problem was studied as part of an early attempt to tackle Hilbert's 21st problem. Consider the following Fuchsian system

$$\frac{d\Psi}{dz} = \sum_{j=1}^{n} \frac{A_j(a)}{\lambda - a_j} \Psi, \tag{4.2}$$

in which we consider the coefficient matrices A_j to be functions of the singular points $a = (a_1, \ldots, a_n)$. Define $A_\infty = -\sum_{j=1}^{n} A_j$. We wish to determine the dependence of the A_js on a such that the monodromy is fixed as the singular points a_k move.

Assume that at each singularity, the eigenvalues of $A_j(a)$ do not differ by integers. Then generically at each singular point these exists an invertible matrix $X_j(a)$ and a diagonal matrix A_j^0 such that

$$A_j^0 = X_j(a)^{-1} A_j X_j(a).$$

Let Ψ be the fundamental matrix of solutions of equation (4.2) normalised at infinity by

$$\Psi(z) = \Phi_\infty(z)z^{-A_\infty^0},$$

where Φ is analytic at infinity and $\Phi_\infty(\infty) = I$.

Now for $j = 1, \ldots, n$, Ψ has the form

$$\Psi = X_j(a)\Phi_j(z)(z - a_j)^{A_j^0} Y_j(a), \tag{4.3}$$

where $\Phi_j(z)$ is analytic at $z = a_j$ and $\Phi_j(a_j) = I$. As Ψ is continued along a small loop γ_j around $z = a_j$, we have

$$(z - a_j)^{A_j^0} = \exp\left(A_j^0 \ln(z - a_j)\right) \mapsto \exp\left(A_j^0[\ln(z - a_j) + 2\pi\mathrm{i}]\right)$$
$$= \exp\left(2\pi\mathrm{i}A_j^0\right)(z - a_j)^{A_j^0}.$$

Hence, after analytic continuation around the loop γ_j, Ψ returns to the value

$$\hat{\Psi} = X_j(a)\Phi_j(z)\exp\left(2\pi\mathrm{i}A_j^0\right)(z - a_j)^{A_j^0}Y_j(a) = \Psi M_j,$$

where

$$M_j = Y_j^{-1}\exp\left(2\pi\mathrm{i}A_j^0\right)Y_j$$

is the corresponding monodromy matrix.

So as the fundamental matrix of solutions Ψ is analytically continued around γ_j, we have $\Psi \mapsto \hat{\Psi} = \Psi M_j$. Now if we impose the isomonodromy condition, that M_j does not change as we vary a, we have $\partial\Psi/\partial a_j \mapsto \hat{\Psi}_{a_j} = (\partial\Psi/\partial a_j)M_j$. It follows that $\Psi_{a_j}(z)\Psi(z)^{-1}$ is a single-valued function of z. It is analytic everywhere except at $a_1, \ldots, a_n$ and infinity, where it has poles. Hence $\Psi_{a_j}(z)\Psi(z)^{-1}$ is a rational function. Now

$$\frac{\partial\Psi}{\partial a_j}\Psi^{-1} = \begin{cases} -\dfrac{A_j}{z - a_j} + O(1), & z \to a_j; \\[2mm] O(1) & z \to a_k, \quad k \neq j; \\[2mm] O(1/z) & z \to \infty. \end{cases}$$

Therefore, for $j = 1, \ldots, n$, the isomonodromy condition for equation (4.2) is

$$\frac{\partial\Psi}{\partial a_j} = -\frac{A_j(a)}{z - a_j}\Psi. \tag{4.4}$$

4.2. The Schlesinger equations

Differentiating equation (4.2) with respect to a_j and using equation (4.4) to eliminate the resulting derivative of Ψ with respect to a_j gives

$$\frac{\partial}{\partial a_j}\left(\frac{\partial\Psi}{\partial z}\right) = \left(\frac{A_j}{(z - a_j)^2} + \sum_{k=1}^{n}\frac{\partial A_k/\partial a_j}{z - a_k} - \sum_{k=1}^{n}\frac{A_k}{z - a_k}\frac{A_j}{z - a_j}\right)\Psi.$$

Similarly differentiating equation (4.4) with respect to z and then using equation (4.2) gives

$$\frac{\partial}{\partial z}\left(\frac{\partial \Psi}{\partial a_j}\right) = \left(\frac{A_j}{(z - a_j)^2} - \sum_{k=1}^{n} \frac{A_j}{z - a_j} \frac{A_k}{z - a_k}\right)\Psi.$$

The equality of the mixed partial derivatives $(\Psi_z)_{a_j} = (\Psi_{a_j})_z$ then results in the *Schlesinger equations:*

$$\frac{\partial A_k}{\partial a_j} = \frac{[A_j, A_k]}{a_j - a_k}, \quad j \neq k;$$

$$\frac{\partial A_k}{\partial a_k} = -\sum_{\substack{j=1 \\ j \neq k}}^{k} \frac{[A_j, A_k]}{a_j - a_k}. \tag{4.5}$$

It can be verified directly that equations (4.4) are compatible for different values of j.

In the case of four Fuchsian singular points, a Möbius transformation can be used to map three of the singularities to 0, 1 and ∞. Call the remaining singularity t. Then the pair of equations (4.2) and (4.4) have the form

$$\frac{\partial \Psi}{\partial z} = \left(\frac{A_0}{z} + \frac{A_1}{z - 1} + \frac{A_t}{z - t}\right)\Psi,$$

$$\frac{\partial \Psi}{\partial t} = -\frac{A_t}{z - t}\Psi. \tag{4.6}$$

The compatibility of this pair of equations gives

$$\frac{dA_0}{dt} = \frac{1}{t}[A_t, A_0], \quad \frac{dA_1}{dt} = \frac{1}{t - 1}[A_t, A_1],$$

$$\frac{dA_t}{dt} = \frac{1}{t}[A_0, A_t] + \frac{1}{t - 1}[A_1, A_t]. \tag{4.7}$$

Without loss of generality we take the matrices A_0, A_1 and A_t to be trace-free (i.e. $\text{Tr}(A_0) = \text{Tr}(A_1) = \text{Tr}(A_t) = 0$). The eigenvalues of A_0, A_1 and A_t (call them $\pm\theta_0$, $\pm\theta_1$ and $\pm\theta_t$) are constants. Furthermore, we use the normalisation

$$A_0 + A_1 + A_t = \begin{pmatrix} -\theta_\infty & 0 \\ 0 & \theta_\infty \end{pmatrix},$$

where θ_∞ is a constant. Let x and y be the $(1,2)$ components of A_0 and A_t respectively. Then

$$u(t) := \frac{tx}{x - (t-1)y}$$

solves the sixth Painlevé equation:

$$\frac{d^2u}{dt^2} = \frac{1}{2}\left\{\frac{1}{u} + \frac{1}{u-1} + \frac{1}{u-t}\right\}\left(\frac{du}{dt}\right)^2 - \left\{\frac{1}{t} + \frac{1}{u-1} + \frac{1}{u-t}\right\}\frac{du}{dt}$$

$$+ \frac{u(u-1)(u-t)}{t^2(t-1)^2}\left\{\alpha + \frac{\beta t}{u^2} + \frac{\gamma(t-1)}{(u-1)^2} + \frac{\delta t(t-1)}{(u-t)^2}\right\}, \qquad (4.8)$$

where

$$\alpha = \tfrac{1}{2}(2\theta_\infty - 1)^2, \quad \beta = -2\theta_0^2, \quad \gamma = 2\theta_1^2, \quad \delta = \tfrac{1}{2}(1 - 4\theta_t^2).$$

The Painlevé equations are six classically known equations that play an important role in the theory of integrable systems. They often appear as reductions of soliton equations. The five other Painlevé equations can be obtained from the sixth Painlevé equation (4.8) by taking appropriate limits.

4.3. *Exercise*

(1) Let A_j, $j = 1, \ldots, n$, be a solution of the Schlesinger equations (4.5). Show that for any positive integer l, the trace of the lth power of A_j (i.e. $\mathrm{Tr}((A_j)^l)$ is a constant for every $j \in \{1, \ldots, n-1\}$. You might want to use the fact that for $m \times m$ matrices X and Y, $\mathrm{Tr}(XY) = \mathrm{Tr}(YX)$.

5. Further Reading

Perhaps the most famous classical text on special functions is [6]. A much more up-to-date monograph is [1], which includes, among other topics, a lot of material on q-special functions. The main classical texts describing the Painlevé equations are [4, 5]. For the connection between the Painlevé and Schlesinger equations and isomonodromy problems, see [3, 7]. In [3], the isomonodromy problem is studied from the point of view of Riemann–Hilbert problems (the problem of determining sectionally analytic functions subject to certain jump conditions). This leads to detailed information about the solutions of the Painlevé equations. Bolibruch's counterexample to Hilbert's 21st problem can be found in [2].

The Painlevé equations and their discrete analogues are still being studied very intensely. Chapter 32 by P. A. Clarkson of the NIST Digital Library of Mathematical Functions is devoted to the solutions of the Painlevé equations: http://dlmf.nist.gov/32

References

[1] G. E. Andrews, R. Askey and R. Roy, *Special Functions*. Encyclopedia of Mathematics and its Applications. Cambridge University Press, Cambridge (2001).

[2] D. V. Anosov and A. A. Bolibruch, *The Riemann–Hilbert Problem*. Aspects of Mathematics, Vol. E22. Friedr. Vieweg and Sohn, Braunschweig (1994).

[3] A. S. Fokas, A. R. Its, A. Kapaev and V. Yu. Novokshenov, *Painlevé Transcendents: The Riemann–Hilbert Approach*. American Mathematical Society, Providence, RI (2006).

[4] E. Hille, *Ordinary Differential Equations in the Complex Domain*. Dover Publications, New York (1997).

[5] E. L. Ince, *Ordinary Differential Equations*. Dover Publications, New York (1944).

[6] E. T. Whittaker and G. N. Watson, *A Course of Modern Analysis*. Cambridge University Press, Cambridge (1927).

[7] H. Żołądek, *The Monodromy Group*. Birkhäuser, Basel (2006).

Solutions and hints for selected exercises

Section 2

(1) $g(z) := 1/(f(z)-1)$ is an entire elliptic function and therefore bounded.

(2) Multiply both sides by y' and integrate.

(4) Multiply by y', integrate and then let $y(z) = au(z) + b$. Choose a and b so that the equation for u has the same form as that for sn.

(6) Use a Möbius transformation to map -1, 0, 2 and 3 to 1, -1, k and $-k$ for some k.

Section 3

(1) and (2) Compare series expansions.

(3) Let $x = 2z$.

Section 4

(1) Use the Schlesinger equations to show that the derivative of $(A_j)^l$ is a sum of commutators.

Chapter 5

Non-commutative Differential Geometry

Shahn Majid

School of Mathematical Sciences,
Queen Mary University of London, London, E1 4NS, UK
s_majid@qmul.ac.uk

Non-commutative geometry is the idea that when geometry is done in terms of coordinate algebras, one does not really need the algebra to be commutative. We provide an introduction to the relevant mathematics from a constructive "differential algebra" point of view that works over general fields and includes the non-commutative geometry of quantum groups as well as of finite groups. We also mention applications to models of quantum spacetime.

1. Converting Geometry to Algebra

Non-commutative geometry of any flavour entails replacing a space and geometric structures on it by an algebra with structures on that, inspired by a precise dictionary such as the one shown in the table below. The dictionary is a crutch which we eventually have to discard as we extend the structures on the algebra so as to make sense even when our algebra is non-commutative. The result is a more general conception of geometry that can even be useful when our algebra is in fact commutative. For example, non-commutative differential structures on a finite set correspond to directed graphs with the given set as vertices. Differentials here do not necessarily commute with functions even though the latter commute amongst themselves. In fact the only commutative calculus here is the zero calculus, so the above is only possible within non-commutative geometry.

Geometry X	Commutative algebra A
polynomial subset $\subseteq \mathbb{C}^n$	reduced finitely generated algebra
compact Hausdorff space	unital (commutative) C^*-algebra
vector bundle	finitely generated projective module
exterior differential forms	differential graded algebra
Dirac operator	spectral triple
group	Hopf algebra coproduct Δ
principal bundle	Hopf–Galois extension
Defined by the algebra	Allow A non-commutative

Of the different approaches, our *quantum groups approach* will be one where we build up the different layers of geometry constructively, guided by group or quantum group symmetries. The latter already revolutionised knot theory and Lie theory through the constructions of knot invariants and canonical bases respectively, so we also want to include these key examples in the same way as classical differential geometry includes Lie groups and their homogeneous spaces. This is in contrast to the well-known operator algebras approach of Connes [1] which comes out of KO-homology and the Dirac operator as the starting point but does not fit too well with the geometry of quantum groups (in Connes approach the most famous example is the "non-commutative torus" which turns out not to be a quantum group). Our approach is also motivated by potential applications to physics through the *quantum spacetime hypothesis* that our actual spacetime coordinates do not commute due to quantum gravity effects.

These lectures will assume an elementary knowledge of algebra. Suffice it to recall that an algebra A over a field k means that A is a vector space over k equipped with a linear map $A \otimes A \to A$ defining an associative product, and is unital if there is an element $1 \in A$ which is the identity for the product. We also recall that a non-zero element of an algebra is called *nilpotent* if some power of it vanishes. An algebra is said to be *reduced* if it has no nilpotent elements. Next we recall that one can think of many geometrical spaces as defined by equations. For example, we can think of a sphere as the set of solutions of $x_1^2 + x_2^2 + x_3^2 = 1$ in $\mathbb{R}^3$. Such sets are called "polynomial" because they are defined as the common zero set of one or more polynomials. We see that the geometry is encoded in an algebra with, in this example, three generators and one relation. Such a conversion to algebra becomes a correspondence if we work over an algebraically closed field such as $\mathbb{C}$.

Theorem 1.1. *There is a one-to-one correspondence between reduced commutative algebras over $\mathbb{C}$ with n generators and polynomial subsets of $\mathbb{C}^n$.*

Proof. This is basically the starting point of algebraic geometry. Recall that an *ideal* $I \subseteq A$ is a linear subspace for which $I.A, A.I \subseteq I$, in which case A/I inherits an algebra structure $(a + I)(b + I) = ab + I$. For any ideal I, its *radical* $\mathrm{rad}(I) = \{a \in A \mid a^n \in I,$ for some $n \in \mathbb{N}\} \supseteq I$ is clearly an ideal obeying $\mathrm{rad}(\mathrm{rad}(I)) = \mathrm{rad}(I)$. It is easy to see that A/I is reduced if and only if $I = \mathrm{rad}(I)$. By definition, a commutative algebra with generators $x_1, \ldots, x_n$, say, means a quotient of the polynomial algebra $\mathbb{C}[x_1, \ldots, x_n]$ by some ideal I (generated by polynomials in the x_i that are set to zero in defining the relations of the algebra). So, reduced algebras of this type are in correspondence with ideals $I \subseteq \mathbb{C}[x_1, \ldots, x_n]$ such that $I = \mathrm{rad}(I)$. Next, any ideal $J \subseteq \mathbb{C}[x_1, \ldots, x_n]$ has a *zero set*

$$Z(J) = \{x \in \mathbb{C}^n \mid a(x) = 0, \ \forall a \in J\}.$$

A polynomial subset $X \subseteq \mathbb{C}^n$ precisely means that $X = Z(J)$ of some ideal J (necessarily generated by a finite collection of polynomials) and Hilbert's "nullstellensatz" says that

$$\mathrm{rad}(J) = \{a \in \mathbb{C}[x_1, \ldots, x_n] \mid a \text{ vanishes on } Z(J)\},$$

which implies that $Z(\mathrm{rad}(J)) = \{x \in \mathbb{C}^n \mid a(x) = 0, \ \forall a \in \mathrm{rad}(J)\} \supseteq Z(J)$. Clearly, $Z(\mathrm{rad}(J)) \subseteq Z(J)$ from the definition of radical, so $Z(\mathrm{rad}(J)) = Z(J)$. We define the corresponding reduced algebra as $\mathbb{C}[x_1, \ldots, x_n]/\mathrm{rad}(J)$ and conversely, given an ideal $I = \mathrm{rad}(I) \subseteq \mathbb{C}[x_1, \ldots, x_n]$, we define a polynomial subset $Z(I)$. These are now clearly inverse. $\qquad\square$

This correspondence is functorial in the sense that appropriately defined maps between objects on either side correspond contravariantly. Motivated by the dictionary, we can now allow our algebras to be non-commutative. Thus we consider any reduced finitely generated algebra over an algebraically closed field as some kind of "non-commutative algebraic space".

Example 1.2. Let $q \in \mathbb{C}^*$. The algebra of 2×2 "quantum matrices" is $\mathbb{C}_q[M_2] = \mathbb{C}\langle a, b, c, d\rangle/I$ where I is the ideal generated by the relations

$$ba = qab, \ ca = qac, \ db = qbd, \ dc = qcd, \ da - ad = (q - q^{-1})bc, \ cb = bc$$

Its further quotient by $ad - q^{-1}bc = 1$ is the "quantum group" $\mathbb{C}_q[\mathrm{SL}_2]$. When $q = 1$, these correspond to $\mathbb{C}^4$ and $\mathrm{SL}_2(\mathbb{C}) \subset \mathbb{C}^4$ respectively.

The correspondence does not work over $\mathbb{R}$ because $\mathbb{R}$ is not algebraically closed. Our approach to this is to work with polynomial subsets of $\mathbb{C}^n$ as the corresponding algebra over $\mathbb{C}$ and specify a "real form" of the subset by means of an antilinear involution $* : A \to A$ with $*^2 = $ id and $(ab)^* = b^*a^*$ for all $a, b \in A$. A map between such $*$-algebras is an algebra homomorphism respecting the $*$ on each side. As an example, if $S^2_{\mathbb{C}}$ is defined by the algebra $\mathbb{C}[x_1, x_2, x_3]$ modulo relations $x_1^2 + x_2^2 + x_3^2 = 1$, the additional information $x_i^* = x_i$ for $i = 1, 2, 3$ picks out the "real sphere" as obtained if we wished to solve the sphere equation in $\mathbb{C}$ with $*$ interpreted as complex conjugation. However, we do not actually have to solve for any such points and when the algebra is non-commutative we take the choice of a $*$ as the definition of a "real form" of the otherwise complex non-commutative space. This is a relatively unexplored field of "$*$-algebraic geometry".

Example 1.3. When q is real we denote by $\mathbb{C}_q[\mathrm{SU}_2]$ the algebra $\mathbb{C}_q[\mathrm{SL}_2]$ in Example 1.2 equipped with the $*$-structure $a^* = d$ and $b^* = -q^{-1}c$. When $q = 1$ the subset of $\mathrm{SL}_2(\mathbb{C}) \subset \mathbb{C}^4$ for which $a^* = \bar{a}, b^* = \bar{b}, c^* = \bar{c}, d^* = \bar{d}$ is the set $\mathrm{SU}_2(\mathbb{C})$ of 2×2 unitary matrices of determinant 1.

More generally, a set X is a topological space if it is equipped with a collection τ of subsets (called "open") obeying some obvious axioms. The familiar case is when X is equipped with "distance" function" $d(x, y)$ with non-negative real values, vanishing if and only if $x = y$, symmetric and obeying the triangle inequality. Then $U \subseteq X$ is open if for every $x \in U$ one can find $\epsilon > 0$ so that the entire ball of points of distance less than ϵ from x lies inside U. The topology in this case is *Hausdorff* in the sense that any two distinct points can be placed in disjoint open sets, while the closure $\overline{X}$ of $X \subset \mathbb{R}^n$ means all points y with the property that a ball of any arbitrarily small size about y intersects X. An *open cover* of X means a collection $\{U_\alpha\}$ of open sets such that $X = \bigcup_\alpha U_\alpha$, and a topological space is *compact* if every open cover has a finite subcover. On the algebra side, a C^*-*algebra* means a $*$-algebra over $\mathbb{C}$ equipped with a norm $|| \, || : A \to \mathbb{R}_{\geq 0}$ (which defines a distance by $d(a, b) = ||a - b||$ and hence a topology on A) subject to some axioms, notably

$$||a^*a|| = ||a||^2, \quad \forall a \in A. \tag{1.1}$$

See elsewhere in this volume for more details. If X is a compact Hausdorff space then the algebra $C(X)$ of complex-valued continuous functions on X

and pointwise operations is a commutative unital C^*-algebra with

$$a^*(x) = \overline{a(x)}, \ \forall x \in X, \quad ||a|| = \sup_{x \in X} |a(x)|.$$

The first theorem of Gelfand and Naimark (1945) says that there is in fact a one-to-one correspondence between unital commutative C^*-algebras and compact Hausdorff topological spaces. This is the second line in our table. Now, according to our philosophy we could regard a non-commutative unital C^*-algebras as a non-commutative analogue of a compact Hausdorff space. The key example is the algebra $B(\mathcal{H})$ of bounded operators on a Hilbert space $\mathcal{H}$, as a non-commutative C^*-algebra with sum and composition of operators,

$$a^* = a^\dagger, \quad ||a|| = \sup_{v \in \mathcal{H}, \, ||v||=1} ||a.v||, \quad \forall a \in B(\mathcal{H}),$$

where † refers to the Hermitian conjugate with respect to the Hilbert space inner product, i.e. $(v, a^\dagger w) = (av, w)$ for all $v, w \in \mathcal{H}$, and the stated norm is the operator norm on $B(\mathcal{H})$. The second Gelfand–Naimark theorem (1954) tells us that any C^*-algebra can be realised as a norm-closed $*$-subalgebra $A \subseteq B(\mathcal{H})$. Just as a space in geometry often can be visualised as embedded in some flat space $\mathbb{R}^n$, a "non-commutative space" in the sense of a C^*-algebra can still be visualised as embedded in $B(\mathcal{H})$. If, as in many treatments of quantum mechanics, one only works with $B(\mathcal{H})$, this would be like only working with $\mathbb{R}^n$ and missing all the non-trivial geometry from the embedding. As a corollary, if we already have a $*$-algebra $\mathcal{A}$, we have only to find a faithful representation $\pi : \mathcal{A} \to B(\mathcal{H})$ as a $*$-algebra map and can then extend $\mathcal{A}$ to a C^*-algebra A as the norm closure $\overline{\pi(\mathcal{A})}$. The (image of) $\mathcal{A}$ appears as a dense $*$-subalgebra and one can say loosely (or more abstractly) that $A = \overline{\mathcal{A}}$, a C^*-algebra "completion" of the $*$-algebra.

Example 1.4. Let $\theta \in [0, 2\pi)$ and let $\mathcal{A}_\theta$ be the $*$-algebra generated by invertible u, v with relations $vu = e^{i\theta}uv$ and $u^* = u^{-1}$, $v^* = v^{-1}$. Let

$$\mathcal{H} = l^2(\mathbb{Z}) = \left\{ (a_n \in \mathbb{C}) \,\Big|\, \sum_n |a_n|^2 < \infty \right\}, \quad (a, b) = \sum_{n \in \mathbb{Z}} \overline{a_n} b_n.$$

This is a Hilbert space on which $\mathcal{A}_\theta$ is represented by

$$(\pi(u)a)_n = a_n e^{in\theta}, \quad (\pi(v)a)_n = a_{n+1}$$

and the *non-commutative torus* A_θ is the resulting C^*-algebra completion of $\mathcal{A}_\theta$. Here $(\pi(u)\pi(v)a)_n = (\pi(v)a)_n e^{in\theta} = a_{n+1}e^{i\theta n}$ while $(\pi(v)\pi(u)a)_n = (\pi(u)a)_{n+1} = e^{i\theta(n+1)}a_{n+1}$, so we indeed have a representation of the algebra $\mathcal{A}_\theta$. Also, if $\theta = 0$ the C^*-algebra is commutative and hence iso-morphic to $C(X)$ for some compact Hausdorff space X given by the set of $*$-algebra homomorphisms. These are necessarily of the form $\phi(u) = e^{\imath\phi_1}$, $\phi(v) = e^{\imath\phi_2}$ for a pair of angles (ϕ_1, ϕ_2) so that $X = S^1 \times S^1$, a classical torus.

2. Quantum Groups

So far we have not yet captured any geometry, only topology. The first "geometrical" structure we look at is a group law on the "underlying space" but in a way that makes sense even when the latter does not exist. Given that our functor in the classical case between geometry and algebra was contravariant, this should appear on the algebra A as a *coalgebra*. Here we fix the field k over which we work and think of the algebra product as a linear map $m : A \otimes A \to A$ and the unit element 1_A equivalently as a linear map $\eta : k \to A$ given by $\eta(1) = 1_A$. A coalgebra consists of the same data obeying the same axioms in an arrow reversed form when written out as commutative diagrams. So, there is a coproduct $\Delta : A \to A \otimes A$ and a counit $\epsilon : A \to k$ obeying coassociativity and counity,

$$(\Delta \otimes \mathrm{id})\Delta = (\mathrm{id} \otimes \Delta)\Delta, \quad (\epsilon \otimes \mathrm{id})\Delta = (\mathrm{id} \otimes \epsilon)\Delta = \mathrm{id}.$$

If Δ were to correspond to an actual product map on an underlying space and ϵ to an identity element then these would be (unital) algebra homomorphisms, where $A \otimes A$ has the tensor product algebra. And if the space was actually a group then there would be an operation $S : A \to A$ induced by inversion. Thus a quantum group or *Hopf algebra* is an algebra A which is also a coalgebra, with Δ, ϵ unital algebra maps and an *antipode* S obeying

$$\cdot(\mathrm{id} \otimes S)\Delta = \cdot(S \otimes \mathrm{id})\Delta = 1\epsilon.$$

It can be shown that S is necessarily antimultiplicative. It should be clear that these axioms, when written as diagrams, are invariant under arrow reversal. That means that for every Hopf algebra construction built from the composition of these structure maps there is a dual "co" construction with arrows reversed. We refer to our books [2, 3] for more details. We will also need a compact notation for the coproduct, called the "Sweedler notation" where we write $\Delta a = a_{(1)} \otimes a_{(2)}$ for any $a \in A$. The suffices here merely tell us which tensor factor of the output of Δa we refer to, and there is a possible sum of such terms understood. We then extend this notation to include iterated coproducts so

$$(\Delta \otimes \mathrm{id})\Delta a = a_{(1)} \otimes a_{(2)} \otimes a_{(3)} = (\mathrm{id} \otimes \Delta)\Delta a,$$

where we may renumber provided we keep the order of the tensor factors.

It was already observed by the topologist E. Hopf in the 1940s that none of this requires A to be commutative, but it is only in the 1980s that many examples emerged (from mathematical physics) which were truly beyond functions on actual (or algebraic) groups or their duals under arrow reversal, i.e. which were non-commutative and non-cocommutative. These were (1) the quasitriangular quantum groups of V.G. Drinfeld, of which the next example is a coordinate algebra version, and (2) the bicrossproduct quantum groups of my PhD thesis. Both include examples associated to every complex semisimple Lie algebra $\mathfrak{g}$.

Example 2.1. $\mathbb{C}_q[\mathrm{SL}_2]$ in Example 1.2 is a Hopf algebra with

$$\Delta \begin{pmatrix} a & b \\ c & d \end{pmatrix} = \begin{pmatrix} a & b \\ c & d \end{pmatrix} \otimes \begin{pmatrix} a & b \\ c & d \end{pmatrix}, \quad \epsilon \begin{pmatrix} a & b \\ c & d \end{pmatrix} = \begin{pmatrix} 1 & 0 \\ 0 & 1 \end{pmatrix}, \quad S \begin{pmatrix} a & b \\ c & d \end{pmatrix} = \begin{pmatrix} d & -qb \\ -q^{-1}c & a \end{pmatrix}.$$

We used a matrix notation and more explicitly,

$$\Delta a = a \otimes a + b \otimes c, \quad \epsilon(a) = 1, \quad Sa = d$$

and so forth. Because we know that Δ, ϵ are algebra maps and S is an anti-algebra map, it suffices to define them on generators, provided these definitions respect the relations of the algebra, for example

$$\begin{aligned}
\Delta(ba) &= (a \otimes b + b \otimes d)(a \otimes a + b \otimes c) \\
&= a^2 \otimes ba + ab \otimes bc + ba \otimes da + b^2 \otimes dc \\
&= qa^2 \otimes ab + qab \otimes ad + qab \otimes (q - q^{-1})bc + ab \otimes bc + qb^2 \otimes cd \\
&= q(a \otimes a + b \otimes c)(a \otimes b + b \otimes d) = \Delta(qab).
\end{aligned}$$

We define a "real form" of a Hopf algebra over $\mathbb{C}$ or Hopf $*$-algebra as $*$-algebra A which is also a Hopf algebra with

$$\Delta \circ * = (* \otimes *)\Delta, \quad \epsilon \circ * = \overline{\epsilon(\ )}, \quad (S \circ *)^2 = \mathrm{id}.$$

An example is $\mathbb{C}_q[\mathrm{SU}_2]$ as in Example 1.3 as a real form of $\mathbb{C}_q[\mathrm{SL}_2]$ when q is real. When $q \in (0,1]$ this Hopf $*$-algebra has a natural completion to a C^*-algebra $C_q(\mathrm{SU}_2)$. Similarly $\mathbb{C}[u, u^{-1}]$ with $u^* = u^{-1}$ and $\Delta u = u \otimes u$ is a Hopf $*$-algebra $\mathbb{C}[S^1]$ and completes to the C^*-algebra $C(S^1)$.

A left A-module or representation of an algebra A on a vector space V is a map $A \otimes V \to V$ (an "action") obeying axioms which polarise the ones for multiplication. Reversing arrows, a left A-*comodule* for a coalgebra A means a vector space V and a "coaction" $\Delta_L : V \to A \otimes V$ obeying

$$(\mathrm{id} \otimes \Delta_L) \circ \Delta_L = (\Delta \otimes \mathrm{id}) \circ \Delta_L, \quad (\epsilon \otimes \mathrm{id})\Delta_L = \mathrm{id}.$$

We may use a shorthand notation $\Delta_L v = v^{(\bar{1})} \otimes v^{(\bar{2})}$. One similarly has the notion of a right coaction $\Delta_R : V \to V \otimes A$. A *comodule algebra* means an algebra B which is a comodule with coaction an algebra map, usually required to be a $*$-algebra map in $*$-algebra setting.

Example 2.2. Let $q \in \mathbb{C}^*$ and $\mathbb{C}_q[\mathbb{C}^2] = \mathbb{C}\langle x, y\rangle / I$ where I is the ideal generated by the relations $yx = qxy$ (the "quantum plane"). This is a $\mathbb{C}_q[\mathrm{SL}_2]$-module algebra with

$$\Delta_L \begin{pmatrix} x \\ y \end{pmatrix} = \begin{pmatrix} a & b \\ c & d \end{pmatrix} \otimes \begin{pmatrix} x \\ y \end{pmatrix}.$$

Again, we use a matrix notation where $\Delta_L x = a \otimes x + b \otimes y$ etc. We extend to products as an algebra map and can check that this is well defined. When $q = 1$ this coaction corresponds to the matrix action $\mathrm{SL}_2(\mathbb{C}) \times \mathbb{C}^2 \to \mathbb{C}^2$.

Hopf algebras provide a cleaner and more logical way of doing many classical constructions, as well as generalising them to the non-commutative case. Here we look at integration. The reader may know that on a (locally) compact group one has a unique translation-invariant integration.

Definition 2.3. Let A be a Hopf algebra over a field k. A left A-*Hopf module* is a vector space V which is both an A-module and an A-comodule and for which $\Delta_L : V \to A \otimes V$ is a left A-module map, i.e. $\Delta_L(a.v) = (\Delta a).(\Delta_L v)$ for all $a \in A, v \in V$.

Here the dot denotes both the action of A on V and the product in A. An example is A itself with the (co)product supplying the *left regular* (co)action. Also, if V is a left A-comodule then its space of coinvariants is

$$^A V = \{v \in V \mid \Delta_L v = 1 \otimes v\}.$$

Lemma 2.4 (Hopf module lemma). *Let V be a left A-Hopf module. Then $V \cong A \otimes (^A V)$ where the right-hand side has the Hopf module structure of A.*

Proof. In one direction $H \otimes ({}^A V) \to V$ the map is the left action $a \otimes v \mapsto a.v$. In the other direction we provide the map

$$v \mapsto F(v) = v^{(\bar{1})}{}_{(1)} \otimes Sv^{(\bar{1})}{}_{(2)}.v^{(\bar{2})}.$$

Let us first verify that this in fact lands in $A \otimes ({}^A V)$, thus

$$
\begin{aligned}
(\mathrm{id} \otimes \Delta_L)F(v) &= v^{(\bar{1})}{}_{(1)} \otimes (Sv^{(\bar{1})}{}_{(2)})_{(1)} v^{(\bar{2})(\bar{1})} \otimes (Sv^{(\bar{1})}{}_{(2)})_{(2)}.v^{(\bar{2})(\bar{2})} \\
&= v^{(\bar{1})}{}_{(1)} \otimes (Sv^{(\bar{1})}{}_{(2)(2)}) v^{(\bar{2})(\bar{1})} \otimes (Sv^{(\bar{1})}{}_{(2)(1)}).v^{(\bar{2})(\bar{2})} \\
&= v^{(\bar{1})}{}_{(1)(1)} \otimes (Sv^{(\bar{1})}{}_{(1)(2)(2)}) v^{(\bar{1})}{}_{(2)} \otimes (Sv^{(\bar{1})}{}_{(1)(2)(1)}).v^{(\bar{2})} \\
&= v^{(\bar{1})}{}_{(1)} \otimes (Sv^{(\bar{1})}{}_{(3)}) v^{(\bar{1})}{}_{(4)} \otimes (Sv^{(\bar{1})}{}_{(2)}).v^{(\bar{2})} \\
&= v^{(\bar{1})}{}_{(1)} \otimes 1 \otimes Sv^{(\bar{1})}{}_{(2)}.v^{(\bar{2})}
\end{aligned}
$$

so that the second factor of $F(v)$ lives in ${}^A V$. The first equality used the Hopf module property of V, the second anticomultiplicativity of S, the third that Δ_L is a coaction. We then renumber coproducts in order to identify two consecutive ones to cancel by the antipode axiom. The map $v \mapsto F(v)$ is inverse to the first map as $v \mapsto F(v) \mapsto v^{(\bar{1})}{}_{(1)}.(S^{(\bar{1})}{}_{(2)}.v^{(\bar{2})}) = (v^{(\bar{1})}{}_{(1)} Sv^{(\bar{1})}{}_{(2)}).v^{(\bar{2})} = \epsilon(v^{(\bar{1})})v^{(\bar{2})} = v$. Similarly the other way. $\square$

We recall that a Hopf algebra coacts on itself by Δ, say from the right.

Definition 2.5. A (right) invariant integral on a Hopf algebra A means a linear map $A \to k$ such that $(\int \otimes \, \mathrm{id})\Delta = \int \otimes 1$.

We will also need the notion of dual Hopf algebra. Recall that if you have a linear map $V \to W$ then its adjoint is a linear map $W^* \to V^*$. Thus the algebra and coalgebra on A are adjoint to a coalgebra and algebra on $H = A^*$ when A is finite dimensional. From the arrow-reversal symmetry of the axioms of a Hopf algebra, we obtain a Hopf algebra again. More generally, two Hopf algebras are *dually paired* if

$$\langle ab, h \rangle = \sum \langle a, h_{(1)} \rangle \langle b, h_{(2)} \rangle, \quad \langle a, hg \rangle = \sum \langle a_{(1)}, h \rangle \langle a_{(2)}, g \rangle$$

etc., for all $a, b \in A$ and $h, g \in H$.

Theorem 2.6. *A finite-dimensional Hopf algebra has a unique right-invariant integral up to normalisation.*

Proof. We note that $H = A^*$ is canonically an A-Hopf module by

$$a.h = \langle h_{(1)}, Sa \rangle h_{(2)}, \quad \Delta_L h = \sum_i e_i \otimes hf^i,$$

where $\{e_i\}$ is a basis of A with dual basis $\{f^i\}$. Here Δ_L defines by evaluation the right multiplication of H on itself as $\langle h, e_i \rangle g f^i = hg$ (summation understood), while the action is the left coaction of H on itself turned into a right action of A similarly by evaluation, turned into a left action by the antipode S. We verify that the two together provide a Hopf module, i.e.

$$\langle h_{(1)}, Sa \rangle e_i \otimes h_{(2)} f^i = a_{(1)} e_a \otimes \langle (hf^i)_{(1)}, Sa_{(2)} \rangle (hf^i)_{(2)}$$

on inserting the definitions (the left-hand side is $\Delta_L(a.h)$ etc.). To see this we evaluate against $g \in H$. Then the right-hand side becomes

$$\langle g_{(1)}, a_{(1)} \rangle \langle (hg_{(2)})_{(1)}, Sa_{(2)} \rangle (hg_{(2)})_{(2)}$$
$$= \langle g_{(1)}, a_{(1)} \rangle \langle Sg_{(2)} Sh_{(1)}, a_{(2)} \rangle h_{(2)} g_{(3)}$$
$$= \langle g_{(1)} Sg_{(2)}, a_{(1)} \rangle \langle Sh_{(1)}, a_{(2)} \rangle h_{(2)} g_{(3)} = \langle h_{(1)}, Sa \rangle h_{(2)} g$$

which is what the left-hand side of the required identity becomes. Once we have a Hopf module, Lemma 2.4 tells us that $H \cong A \otimes (^A H)$ where

$$^A H = \{h \in H \mid e_i \otimes hf^i = 1 \otimes h\} = \{h \in H \mid e_i \langle hf^i, a \rangle = h(a), \ \forall a \in A\}$$

$$= \{h \in H \mid h(a_{(1)}) a_{(2)} = h(a), \ \forall a \in A\}.$$

We see that this is the space of right-invariant integrals $\int : A \to k$. Hence this space is one-dimensional as A, H have the same dimension.   □

In the infinite-dimensional case one can show similarly that the space of right-invariant integrals has dimension at most 1, i.e. an integral may not exist but if it does, it is unique. If A is a Hopf $*$-algebra, we require $\int(a^*) = \overline{\int a}$ for all $a \in A$.

Example 2.7. On $\mathbb{C}_q[SL_2]$ the "Haar integral" takes the form

$$\int (bc)^n = \frac{(-1)^n q^n}{[n+1]_{q^2}}, \qquad [m]_q = \frac{1 - q^m}{1 - q},$$

and zero on other monomials in a, b, c or in b, c, d. We suppose that q is generic in the sense that all the denominators are invertible. For example,

$$\left(\int \otimes \, \mathrm{id} \right) \Delta(bc) = \left(\int \otimes \, \mathrm{id} \right) ((a \otimes b + b \otimes d)(c \otimes a + d \otimes c))$$

$$= \int ac \otimes ba + \int ad \otimes bc + \int bc \otimes da + \int bd \otimes dc$$

$$= \int (1 + q^{-1} bc + qbc) \otimes bc + \int bc \otimes 1 = \int bc \otimes 1$$

since $\int bc = -1/(q + q^{-1})$.

There are many applications of this theory. Here we mention only one: For any finite-dimensional Hopf algebra A with dual H, one has Fourier transform

$$\mathcal{F} : A \to H, \quad \mathcal{F}(a) = \sum_i \left(\int e_i a \right) f^i,$$

$$\mathcal{F}^{-1}(h) = \frac{1}{\int^*(\int)} \sum_i S^{-1} e_i \int^* h f^i.$$

Here $\int$ is a right-invariant integral, which exists by the above results, and $\int^*$ is similarly one on H. The 2π normalisation factor in usual Fourier theory becomes $\int^*(\int)$. One can show that both this and S are invertible. Usually, Fourier transform is limited to Abelian groups but using Hopf algebra technology it applies more generally. For example, the quantum group $\mathbb{C}_q[\mathrm{SL}_2]$ has a finite-dimensional quotient $c_q[\mathrm{SL}_2]$ when q is a primitive rth root of 1, namely by the additional relations $b^r = c^r = 0$ and $a^r = d^r = 1$. Its dual is the Hopf algebra $u_q(\mathrm{sl}_2)$ and so $\mathcal{F} : c_q[\mathrm{SL}_2] \to u_q(\mathrm{sl}_2)$.

3. Non-commutative Differential Forms

In differential geometry one equips a topological space with the structure of a differentiable manifold. This means that in each open set of a cover, one has "local coordinates" $\{x_i\}$ identifying the open set with a region of $\mathbb{R}^n$ (some fixed n which is the dimension of the manifold) and with the transition functions between different such "patches" being differentiable. One also defines the tangent bundle with sections spanned by vector fields of the form $\sum_i v_i(x) \frac{\partial}{\partial x_i}$ in each local patch. The cotangent bundle is dual to this and the space of "1-forms" Ω^1 is spanned by elements of the form $\sum \omega^i(x) \mathrm{d}x_i$ in each local patch. Here $\mathrm{d}x_i$ are a dual basis to $\frac{\partial}{\partial x_i}$ at each point. One also has an abstract map d which turns a function f into a differential 1-form $\mathrm{d}f = \sum_i \frac{\partial f}{\partial x_i} \mathrm{d}x_i$. We turn all this around and define Ω^1 by its desired properties as a "non-commutative differentiable structure".

3.1. *Differentials on algebras*

Let A be a unital algebra over a field k. A vector space is a *bimodule* over A if it is both a left and a right A-module and if the two actions commute.

Definition 3.1. A first-order "differential calculus" (Ω^1, d) over A means

(1) Ω^1 an A-bimodule;
(2) a linear map $\mathrm{d} : A \to \Omega^1$ (the *exterior derivative*) such that

$$\mathrm{d}(ab) = (\mathrm{d}a)b + a\mathrm{d}b, \quad \forall a, b \in A;$$

(3) $\Omega^1 = \mathrm{span}\{a\mathrm{d}b \mid a, b \in A\}$;
(4) (optional *connectedness condition*) $\ker \mathrm{d} = k.1$.

This is more or less the minimum that one could require for an abstract notion of "differentials" — one should be able to multiply them from the left and right by elements of A and have a Leibniz rule with respect to this. In usual algebraic geometry one would assume that the left and right modules coincide, i.e. that $a\mathrm{d}b = \mathrm{d}b.a$ for all $a, b \in A$, but this is not reasonable to impose when our algebras are non-commutative. For if we did, we would have $\mathrm{d}(ab - ba) = 0$ so that d would be far from connected. Our definition is necessarily more general and interesting even for commutative algebras. In some cases, for example if Ω^1 is free as a left or right A-module, we have a well-defined left or right *cotangent dimension*.

Proposition 3.2. *Every A has a universal differential calculus given by*

(1) $\Omega^1_{\mathrm{univ}} = \ker m \subseteq A \otimes A$ *(the kernel of the product map)*;
(2) $\mathrm{d} : A \to \Omega^1_{\mathrm{univ}}$ *is given by* $\mathrm{d}a = 1 \otimes a - a \otimes 1$ *for* $a \in A$.

The calculus is connected. Any other differential calculus on A is a quotient $\Omega^1 = \Omega^1_{\mathrm{univ}}/\mathcal{N}$, for some subbimodule $\mathcal{N} \subseteq \Omega^1_{\mathrm{univ}}$. Its exterior derivative is that of Ω^1_{univ} followed by the projection to the quotient.

Proof. It is elementary to check that Ω^1_{univ} is indeed a differential calculus. To see that it obeys axiom 4 (is connected), suppose that A' is a chosen complement to $k1$ so that $A = k1 \oplus A'$. If $a = \lambda 1 + b \in k1 \oplus A'$ then $\mathrm{d}a = 1 \otimes b - b \otimes 1$. Hence if $\mathrm{d}a = 0$ we see that $1 \otimes b = b \otimes 1$. By projecting the first factor onto A' we conclude that $b = 0$ as required. The universal property follows from the surjectivity axiom 3. $\qquad\square$

Note that universal here means with the obvious notion of morphisms between calculi, namely bimodule maps that form a commutative triangle with the exterior derivatives.

Example 3.3. Let X be a finite set and $A = \mathbb{C}(X)$ the algebra of functions on it. (Connected) Ω^1 on $\mathbb{C}(X)$ are in one-to-one correspondence with (connected) directed graphs Γ with vertices X. We write $x \to y$ for the edges, then $\Omega^1 = \mathrm{span}_{\mathbb{C}}\{e_{x \to y}\}$ with

$$\mathrm{d}f = \sum_{x \to y}(f(y) - f(x))e_{x \to y}, \quad f.e_{x \to y} = f(x)e_{x,y}, \quad e_{x \to y}.f = e_{x \to y}f(y).$$

We can deduce that $e_{x \to y} = \delta_x \mathrm{d}\delta_y$ for all $x \to y$, where δ_x is the Kronecker delta-function. Note that $A \otimes A$ has basis $\{\delta_x \otimes \delta_y\}$ of which Ω^1_{univ} is the subspace with basis restricted to $x \neq y$, and $\mathrm{d}_{\mathrm{univ}}\delta_x = 1 \otimes \delta_x - \delta_x \otimes 1 = \sum_{y \neq x} \delta_y \otimes \delta_x - \delta_x \otimes \delta_y$ which takes the form stated with the maximal graph $x \to y$ for all $x \neq y$. Here $\delta_x \mathrm{d}_{\mathrm{univ}}\delta_y = \delta_x \otimes \delta_y$ when $x \neq y$. Now suppose that we have some other calculus defined by a sub-bimodule $\mathcal{N}$. If $n = \sum_{x \neq y} n_{x,y}\delta_x \otimes \delta_y \in \mathcal{N}$ then $\delta_x n \delta_y = n_{x,y}\delta_x \otimes \delta_y \in \mathcal{N}$. Hence either $n_{x,y} = 0$ for all elements n or $\delta_x \otimes \delta_y \in \mathcal{N}$. Hence $\mathcal{N} = \{\delta_x \otimes \delta_y \mid (x,y) \in \bar{E}\}$ for some subset $\bar{E} \subseteq (X \times X)\backslash\mathrm{diagonal}$. The quotient of the universal calculus by $\mathcal{N}$ can therefore be identified with the subspace spanned by $\delta_x \otimes \delta_y$ for $(x,y) \in E$ where E is the complement of $\bar{E}$ in $(X \times X)\backslash\mathrm{diagonal}$. Such E are the edges of our digraph. Clearly $\ker \mathrm{d}$ consists of those functions for which $f(y) = f(x)$ for all $x \to y$, which are a multiple of 1 if and only if the graph is connected. If the graph is regular in the sense that out of every vertex there are a fixed number n of edges then X has left contangent dimension n and is, moreover, parallelisable in the sense $\Omega^1 \cong \mathbb{C}(X) \otimes \mathbb{C}^n$.

Definition 3.4. An *exterior algebra* on A or "differential graded algebra" means

(1) a graded algebra $\Omega = \bigoplus_n \Omega^n$ with $\Omega^0 = A$,
(2) $\mathrm{d} : \Omega^n \to \Omega^{n+1}$ such that $\mathrm{d}^2 = 0$ and

$$\mathrm{d}(\omega\rho) = (\mathrm{d}\omega)\rho + (-1)^n\omega\mathrm{d}\rho, \quad \forall \omega, \rho \in \Omega, \; \omega \in \Omega^n,$$

(3) A, Ω^1 generate Ω.

Its *non-commutative de Rham cohomology* is defined to be

$$H^n(A) = \ker(\mathrm{d}|_{\Omega^n})/\mathrm{image}(\mathrm{d}|_{\Omega^{n-1}}).$$

In the last definition we understand $\mathrm{d}_{\Omega^n} = 0$ if $n < 0$. The *volume dimension* (if it exists) is the top degree and can be different from the cotangent dimension, if that exists. In practice one can just construct Ω up to and including any degree of interest and "fill in" all higher degrees automatically. For example, every first-order calculus Ω^1 on A has a *maximal prolongation* to an exterior algebra generated by these with the minimal further relations contained in the definition of an exterior algebra.

Theorem 3.5. *Let A be an algebra. The maximal prolongation of its universal first-order calculus is its universal exterior algebra $\Omega_{\mathrm{univ}} = \bigoplus_n \Omega_{\mathrm{univ}}^n$ where $\Omega_{\mathrm{univ}}^n \subset A^{\otimes(n+1)}$ is the joint kernel of all the product maps between adjacent copies of A in the tensor product. The product and differential are*

$$(a_0 \otimes \cdots \otimes a_n)(b_0 \otimes \cdots \otimes b_m) = (a_0 \otimes \cdots \otimes a_n b_0 \otimes \cdots \otimes b_m),$$

$$d(a_0 \otimes \cdots \otimes a_n) = \sum_{i=0}^{n+1} (-1)^i a_0 \otimes \cdots \otimes a_{i-1} \otimes 1 \otimes a_i \otimes \cdots \otimes a_n.$$

Moreover, $H^i(A) = k$ if $i = 0$ and is zero otherwise (the complex is acyclic).

Proof. We leave the reader to verify that this is the maximal prolongation and focus on the cohomology. Consider

$$\omega = \sum_\alpha a_0^\alpha da_1^\alpha \cdots da_n^\alpha = \sum_\alpha \lambda^\alpha da_1^\alpha \cdots da_n^\alpha + \sum_\alpha b^\alpha da_1^\alpha \cdots da_n^\alpha,$$

where we replace $a_0^\alpha = \lambda^\alpha 1 \oplus b^\alpha$ according to $A = k1 \oplus A'$ (as before). Then $d\omega = \sum_\alpha db^\alpha da_1^\alpha \cdots da_n^\alpha$ as $d1 = 0$ and $d^2 = 0$. Hence if ω is closed, and writing out db^α, we see that

$$1 \otimes \sum_\alpha b^\alpha da_1^\alpha \cdots da_n^\alpha = \sum_\alpha b^\alpha \otimes da_1^\alpha \cdots da_n^\alpha.$$

Now projecting the first factor to A' we conclude that in this case $\sum_\alpha b^\alpha da_1^\alpha \cdots da_n^\alpha = 0$ and hence

$$\omega = \sum_\alpha \lambda^\alpha da_1^\alpha \cdots da_n^\alpha = d\left(\sum_\alpha \lambda^\alpha a_1^\alpha da_2^\alpha \cdots da_n^\alpha\right)$$

so that ω is exact. The last step requires $n > 1$. The $n = 0$ case was already dealt with and we computed $\ker d = k1$ in this case. $\square$

We see that the universal calculus is both too big and has no non-trivial "topology" in view of this theorem. For a more non-trivial example of the same dimensions as the classical geometry, one has the following example.

Example 3.6. On the algebraic non-commutative torus $\mathcal{A}_\theta$ we let $\Omega^1 = \mathcal{A}_\theta\{du, dv\}$ with

$$du.u = u.du, \quad dv.v = v.dv, \quad dv.u = e^{i\theta} u.dv, \quad du.v = e^{-i\theta} v.du.$$

Its maximal prolongation has $\Omega^2 = \mathcal{A}_\theta du dv$ and $\Omega^i = 0$ for all $i > 2$. The non-commutative de Rham cohomology is the same as classically, namely

$$H^0 = \mathbb{C}, \quad H^1 = \mathbb{C}^2, \quad H^2 = \mathbb{C}.$$

Here u, u^{-1} generate a classical circle $\mathbb{C}[S^1]$ and Ω^1 restricts to this as classically. Similarly for v, v^{-1}. Of interest is how the two interact. From d applied to the relations we need $d(vu) = dv.u + v.du = e^{i\theta} d(uv) = e^{i\theta} du.v + e^{i\theta} u.dv$, which is provided by the relations shown. These relations also ensure that the calculus becomes the usual one on $S^1 \times S^1$ when $\theta = 0$. We thus define Ω^1 as a free module on the left (i.e. just by the product of $\mathcal{A}_\theta$) and use the relations between 1-forms and the algebra generators to define the right module structure. Next, apply d to these relations to find

$$(du)^2 = 0, \quad (dv)^2 = 0, \quad dvdu + e^{i\theta} dudv = 0.$$

which tells us that Ω^2 is one-dimensional over $\mathcal{A}_\theta$. That the cohomology is the same as classically is a calculation along the same lines as when $\theta = 0$. One still has a basis $\{u^m v^n \; : \; m, n \in \mathbb{Z}\}$ of $\mathcal{A}_\theta$ and there are extra $e^{i\theta}$ factors in the computations but the steps are the same. The natural basis for computations is given by $e_1 = u^{-1} du$ and $e_2 = v^{-1} dv$ with $de_1 = -u^{-1} du.u^{-1} du = 0$ from the Leibniz rule and the commutation relations above, as classically. Similarly for de_2. Now suppose that

$$\lambda e_1 + \mu e_2 = d\left(\sum a_{mn} u^m v^n\right) = \sum a_{mn}(m u^{m-1} du.v^n + u^m n v^{n-1} dv).$$

as d on powers of u alone or v alone behaves as classically. Putting du to the right introduces a factor $e^{-ni\theta}$ but we still need $a_{mn} = 0$ for all $m \neq 0$ or $n \neq 0$ for the du and dv coefficients to match. Hence only $a_{0,0}$ can contribute and this does not as $d1 = 0$. Hence all $\lambda e_1 + \mu e_2$ are not exact. In fact, $H^0 = \mathbb{C}.1$, $H^1 = \mathbb{C}e_1 \oplus \mathbb{C}e_2$ and $H^2 = \mathbb{C}e_1 e_2$.

Definition 3.7. If A is a $*$-algebra we ask the calculus or exterior algebra to be $*$-compatible in the sense that this extends to Ω a graded-$*$-algebra (i.e. with an extra sign factor according to the degrees) with $[*, d] = 0$.

Examples are the universal calculus on any $*$-algebra and the above calculus on $\mathcal{A}_\theta$, where $e_i^* = -e_i^*$ in keeping with the classical picture $u = e^{i\phi_1}, v = e^{i\phi_2}$ where $e_i = id\phi_i$.

3.2.   *Differentials on quantum groups*

In the classical case of a Lie group one has a unique translation-covariant differential calculus. In the quantum group we again have relatively few such calculi but usually without uniqueness.

Definition 3.8. Ω^1 on a Hopf algebra A is *left covariant* if:

(1) There is a left coaction $\Delta_L : \Omega^1 \to A \otimes \Omega^1$;
(2) Ω^1 becomes a left Hopf module;
(3) $\mathrm{d} : A \to \Omega^1$ is a comodule map, where A coacts on itself by Δ.

Note that since Ω^1 is spanned by elements of the form $a\mathrm{d}b$, conditions (2), (3) imply that Δ_L is a bimodule map and defined by the formula

$$\Delta_L(a\mathrm{d}b) = \sum a_{(1)}b_{(1)} \otimes a_{(2)}\mathrm{d}b_{(2)}.$$

Conversely, if Δ_L is well defined by such a formula then it is easy to verify that (1), (2), (3) hold. So this is all we need for a calculus to be left-covariant. The *left-invariant 1-forms* are the coinvariants $\Lambda^1 = {}^A\Omega^1$.

Theorem 3.9. *Let A be a Hopf algebra. Left-covariants Ω^1 on A have the form $\Omega^1 {\cong} A \otimes \Lambda^1$ where $\Lambda^1 {\cong} A^+/I$ for some right ideal $I \subseteq A^+ = \ker \epsilon$.*

Proof.   The first part is the Hopf module lemma, Lemma 2.4. Next, consider the "Maurer–Cartan form" $\omega : A^+ \to \Lambda^1$ defined by

$$\omega(a) = Sa_{(1)}\mathrm{d}a_{(2)}, \quad \forall a \in A^+.$$

Its image is invariant since $\Delta_L\omega(a) = (Sa_{(1)})_{(1)}a_{(2)(1)} \otimes (Sa_{(1)})_{(2)}\mathrm{d}a_{(2)(2)} = (Sa_{(1)(2)})a_{(2)(1)} \otimes Sa_{(1)(1)}\mathrm{d}a_{(2)(2)} = 1 \otimes \omega(a)$. Moreover, if $\alpha = a_i\mathrm{d}b_i$ is left-invariant then $a_{i(1)}b_{i(1)} \otimes a_{i(2)}\mathrm{d}b_{i(2)} = 1 \otimes \alpha$. Applying S to the first factor and multiplying tells us that $\alpha = \omega(\epsilon(a_i)b_i)$. Hence ω is surjective and $\Lambda^1 {\cong} A^+/I$ where $I = \ker \omega$. The Leibniz rule requires I a right ideal. $\square$

The converse is also true, so left-covariant differential calculi (Ω^1, d) are in one-to-one correspondence with right ideals $I \subseteq A^+$. Given an ideal I we have $\Lambda^1 = A^+/I$ as a right H-module by right multiplication in A. The left (co)action on Ω^1 is just the (co)action on A as in the Hopf module lemma. The rest of the structure is just tracing through the above isomorphism. Thus

$$a.(b \otimes v) = ab \otimes v, \quad \Delta_L(a \otimes v) = (\Delta \otimes \mathrm{id})(a \otimes v),$$

$$(a \otimes v).b = \sum ab_{(1)} \otimes v.a_{(2)}$$

where the last of these follows from $\omega(a).b = \sum a_{(1)}\omega(ba_{(2)})$. Similarly,

$$da = (\mathrm{id} \otimes \pi)(\Delta a - a \otimes 1), \quad \pi : A^+ \to \Lambda^1$$

where π is the canonical surjection. $I = \{0\}$ gives the universal calculus.

There is a similar notion of right-covariance and a calculus is called *bicovariant* if it is both left and right covariant. The two coactions will necessarily commute (so that Ω^1 is a bicomodule).

Corollary 3.10. *A left-covariant Ω^1 is bicovariant if and only if $\mathrm{Ad}_R(I) \subseteq I \otimes A$ where $\mathrm{Ad}_R(a) = \sum a_{(2)} \otimes Sa_{(1)}a_{(3)}$. If so then Λ^1 is a right A-comodule from Ad_R.*

Proof. If Δ_R is well defined, we compute

$$\Delta_R(a\omega(b)) = a_{(1)}(Sb_{(1)})_{(1)}db_{(2)(1)} \otimes a_{(2)}(Sb_{(1)})_{(2)}b_{(2)(2)}$$

$$= a_{(1)}Sb_{(2)(1)}db_{(2)(2)} \otimes h_{(2)}(Sb_{(1)})b_{(3)}$$

$$= a_{(1)}\omega(b_{(2)}) \otimes a_{(2)}(Sb_{(1)})b_{(3)}.$$

We recognise the right adjoint coaction $\mathrm{Ad}_R(b)$ which at the classical level corresponds to conjugation in the group, and require that $\mathrm{Ad}_R(I) \subseteq I \otimes A$. Conversely, in this case we define Δ_R as stated. Equivalently, we define the right coaction on $A \otimes \Lambda^1$ as the tensor product of Δ and Ad_R. $\qquad\square$

An advantage of the bicovariant case is that there is a natural prolongation in some sense compatible with Poincaré duality. We define

$$\Psi : \Lambda^1 \otimes \Lambda^1 \to \Lambda^1 \otimes \Lambda^1, \quad \Psi(\omega(a) \otimes \omega(b)) = \omega(b_{(2)}) \otimes \omega(a(Sb_{(1)})b_{(3)})$$

obeying the Yang–Baxter or braid relations. Any permutation $\sigma \in S_n$ can be written in reduced form $\sigma = \sigma_{i_1} \cdots \sigma_{i_{l(\sigma)}}$ where $l(\sigma)$ is the length and $\sigma_i = (i, i+1)$ and we define $\Psi_\sigma = \Psi_{i_1} \cdots \Psi_{i_{l(\sigma)}}$ on $(\Lambda^1)^{\otimes n}$, where Ψ_i denotes the braiding in the $i, i+1$ tensor factors, and "antisymmetrise" by

$$A_n = \sum_\sigma (-1)^{l(\sigma)}\Psi_\sigma, \quad \Lambda^n = \frac{(\Lambda^1)^{\otimes n}}{\ker A_n}, \quad \Omega^n = A \otimes \Lambda^n.$$

This is the original definition of Woronowicz [4] while another approach is due to myself based on braided-integers and Λ as a superbraided Hopf algebra canonically associated to Λ^1. We now focus on bicovariant Ω^1. One says that a calculus is *irreducible* if it has no proper quotients.

Example 3.11. For $A = \mathbb{C}[x]$ with $\Delta x = x \otimes 1 + 1 \otimes x$ (the affine line) the irreducible bicovariant Ω^1 are parametrised by $\lambda \in \mathbb{C}$ and take the form

$$\Omega^1 = \mathbb{C}[x].\mathrm{d}x, \quad \mathrm{d}x.f(x) = f(x+\lambda)\mathrm{d}x, \quad \mathrm{d}f = \frac{f(x+\lambda) - f(x)}{\lambda}\mathrm{d}x.$$

Only the Newton–Leibniz calculus at $\lambda = 0$ has $[\mathrm{d}x, f] = 0$. Here Ad_R is trivial so left covariant calculi are the same as bicovariant ones and $\mathbb{C}[x]$ is a principal ideal domain so $A^+ = \langle x \rangle$ (the ideal generated by x) and $I = \langle xm(x) \rangle$ for some monic irreducible polynomial. Hence calculi are in one-to-one correspondence with such m. Over $\mathbb{C}$, the only possible m are $m(x) = x - \lambda$ for $\lambda \in \mathbb{C}$. It remains only to work out the structure of the calculus in this case. We have $\Lambda^1 = \langle x \rangle / \langle x(x - \lambda) \rangle$ is one-dimensional and

$$\mathrm{d}x = 1 \otimes \pi(x), \quad \mathrm{d}x.x = x_{(1)} \otimes \pi(xx_{(2)}) = x \otimes \pi(x) + 1 \otimes \pi(x^2) = x\mathrm{d}x + \lambda\mathrm{d}x$$

as $x^2 = x(x - \lambda) + \lambda x$. The bimodule relations with general $f(x)$ follow. Also $\mathrm{d}x^n = \mathrm{d}x.x^{n-1} + x\mathrm{d}x^{n-1} = (x+\lambda)^{n-1}\mathrm{d}x + x\mathrm{d}x^{n-1}$ gives the formula for d on monomials by induction.

The first part of the analysis applies to $k[x]$ for any field k; calculi are again classified by monic irreducibles $m(x)$, and Ω^1 as a left $k[x]$-module is identified with $k_\lambda[x]$, where λ generates the field extension defined by $m(\lambda) = 0$.

Example 3.12. For $A = \mathbb{C}[t, t^{-1}]$ with $\Delta t = t \otimes t$ (the complex circle) the irreducible bicovariant Ω^1 are parametrised by $q \in \mathbb{C}^*$ and take the form

$$\Omega^1 = \mathbb{C}[t, t^{-1}].\mathrm{d}t, \quad \mathrm{d}t.f(t) = f(qt)\mathrm{d}t, \quad \mathrm{d}f = \frac{f(qt) - f(t)}{q(t-1)}\mathrm{d}t.$$

Only the Newton–Leibniz calculus at $q = 1$ has $[\mathrm{d}t, f] = 0$. Here $A^+ = \langle t - 1 \rangle$, with the elements given by $t^n - 1 = [n]_t(t - 1)$ in terms of the q-integer $[n]_t = 1 + t + \cdots + t^{n-1}$. As before, an ideal in A^+ takes the form $I = \langle (t-1)(t-q) \rangle$. Then

$$\mathrm{d}t = t \otimes \pi(t - 1), \quad \mathrm{d}t.t = t^2 \otimes \pi((t-1)t) = qt^2 \otimes \pi(t - 1) = qt\mathrm{d}t,$$

where $(t - 1)t = (t - 1)(t - q) + q(t - 1)$. We see that $\Omega^1 = A.e_1$ where $e_1 = t^{-1}\mathrm{d}t$ is the more natural left-invariant generator. We then obtain the bimodule relations with general $f(t)$. Finally,

$$\mathrm{d}t^n = \mathrm{d}t.t^{n-1} + t\mathrm{d}t.t^{n-2} + \cdots + t^{n-1}\mathrm{d}t$$

$$= t^{n-1}(q^{n-1} + \cdots + q + 1)\mathrm{d}t = [n]_q t^{n-1}\mathrm{d}t$$

to give the result stated. It is instructive to compute the cohomology. We assume q is not a non-trivial root of unity, which is equivalent to $[n]_q \neq 0$ for all $n \neq 0$. Then $\mathrm{d}(\sum a_n t^n) = \sum t^{n-1}[n]_q a_n \mathrm{d}t = 0$ implies that $a_n = 0$ for all $n \neq 0$, i.e. $H^0 = \mathbb{C}.1$ and our calculus is connected. Also, $\Omega^i = 0$ for $i > 1$ since in any prolongation we have $0 = \mathrm{d}(\mathrm{d}t.t - qt\mathrm{d}t) = -(\mathrm{d}t)^2(1+q) = -[2]_q(\mathrm{d}t)^2$, and hence $(\mathrm{d}t)^2 = 0$. Then we similarly have

$$\sum a_n t^n \mathrm{d}t = \mathrm{d}\left(\sum \frac{a_n t^{n+1}}{[n+1]_q}\right)$$

if and only if $a_{-1} = 0$, and hence $H^1 = \mathbb{C}t^{-1}\mathrm{d}t$. This changes if q is a root of unity.

Example 3.13. For $A = k(G)$ on a finite group G, the irreducible bicovariant Ω^1 are in one-to-one correspondence with non-trivial conjugacy classes $\mathcal{C} \subset G$. The 1-forms $e_a = \sum_{g \in G} \delta_g \mathrm{d}\delta_{ga}$ for $a \in \mathcal{C}$ form a basis of Λ^1 and

$$e_a.f = R_a(f)e_a, \quad \mathrm{d}f = \sum_{a \in \mathcal{C}}(R_a(f) - f)e_a.$$

The corresponding graph is the Cayley graph of G with respect to $\mathcal{C}$. Here $R_a(f) = f((\ )a)$ is right-translation. For Λ^1 we set to zero all delta-functions except $\{\delta_a\}_{a \in \mathcal{C}}$. These project to our basis $\{e_a\}$, which we identify in terms of d. The calculus is connected if and only if $\mathcal{C}$ is a generating set.

We see that any generating set of a finite group gives a left-covariant connected calculus while ad-stable such sets give a bicovariant one. In the second case we have an exterior algebra $\Omega(G)$. Its volume dimension carries deep information (conjecturally related to Lusztig's canonical basis in the case where G is the Weyl group of a semisimple Lie algebra). Here $G = S_3$ with $\mathcal{C} = \{(12), (13), (23)\}$ has dimensions $\dim(\Omega^i) = 1:3:4:3:1$, so volume dimension 4 but cotangent dimension 3.

Example 3.14. $\mathbb{C}_q[SL_2]$ has a left-covariant Ω^1 (compatible with $*$) with

$$e_- = d\mathrm{d}b - qb\mathrm{d}d, \quad e_+ = q^{-1}a\mathrm{d}c - q^{-2}c\mathrm{d}a, \quad e_0 = d\mathrm{d}a - qb\mathrm{d}c,$$

$$e_{\pm}f = q^{|f|}fe_{\pm}, \quad e_0f = q^{2|f|}fe_0,$$

$$\mathrm{d}\begin{pmatrix} a \\ c \end{pmatrix} = \begin{pmatrix} a \\ c \end{pmatrix}e_0 + q\begin{pmatrix} b \\ d \end{pmatrix}e_+, \quad \mathrm{d}\begin{pmatrix} b \\ d \end{pmatrix} = \begin{pmatrix} a \\ c \end{pmatrix}e_- - q^{-2}\begin{pmatrix} b \\ d \end{pmatrix}e_0.$$

Here $|f|$ of a monomial is the number of a, c minus the number of b, d. As with the non-commutative torus, for generic q the maximal prolongation gives an exterior algebra of classical dimensions, e.g. a unique top form $e_+ e_-$ up to normalisation.

For all standard quantum groups $\mathbb{C}_q[G]$ associated to simple groups, there do *not* exist bicovariant calculi of classical dimensions. For generic q and a $q \to 1$ limiting requirement, one has one calculus for every irreducible representation V of the group, of cotangent dimension $(\dim(V))^2$. For $\mathbb{C}_q[SL_2]$ the smallest bicovariant calculus is therefore four-dimensional with dimension pattern $\dim(\Omega^i) = 1{:}4{:}6{:}4{:}1$ like a 4-manifold. Interestingly, the non-commutative de Rham cohomology is the same for $\mathbb{C}_q[SL_2]$ (or $\mathbb{C}_q[SU_2]$ as a real form) as for the above calculus on S_3, namely

$$H^0 = \mathbb{C}, \quad H^1 = \mathbb{C}, \quad H^2 = 0, \quad H^3 = \mathbb{C}, \quad H^4 = \mathbb{C}.$$

4.   Non-commutative Vector Bundles

Cotangent bundles are special cases of vector bundles. On a manifold X, a vector bundle of rank n means a smooth assignment of a vector space E_x at each point x with each E_x isomorphic to $\mathbb{C}^n$. More precisely, E is itself a manifold and there is a surjection $\pi : E \to X$ such that in an open neighbourhood of any point we have $\pi^{-1}(U) \cong U \times \mathbb{C}^n$ compatible with π and compatible with a vector space structure on each $E_x = \pi^{-1}(x)$. The main thing we will need is that every vector bundle E is the direct summand of a trivial one: there is some bundle $E' \to X$ such that $E \oplus E' \cong X \times \mathbb{C}^N$ for some sufficiently large N. We also need the space of sections

$$\Gamma(E) = \{\text{smooth maps } s : X \to E \mid \pi \circ s = \text{id}\}$$

which is a module over the algebra of functions on X by pointwise multiplication if we think of sections as "functions" with values $s(x) \in \pi^{-1}(x) = E_x$. In terms of sections, $\Gamma(E) \oplus \Gamma(E') \cong C^\infty(X) \otimes \mathbb{C}^N$ and hence $\Gamma(E) = \text{image}(e)$ for some $e \in M_N(C^\infty(X))$ a projection matrix $e^2 = e$. In this way (made precise in the "Serre–Swan theorem") vector bundles become equivalent to finitely generated projective modules over the algebra of functions.

4.1.   *Projective modules and K-theory*

Let A be an algebra over a field k. In view of the above, we have the following definition.

Definition 4.1. A "vector bundle" over A is defined as a finitely generated projective $\mathcal{E}$ module over A.

If Ω^1 is a differential calculus on A and $\mathcal{E}$ a vector bundle on A, we define a connection on $\mathcal{E}$ as a linear map such that

$$D : \mathcal{E} \to \Omega^1 \underset{A}{\otimes} \mathcal{E}, \quad D(as) = \mathrm{d}a \underset{A}{\otimes} s + a Ds, \quad \forall a \in A,\ s \in \mathcal{E}.$$

If higher forms are defined we extend the definition to "form-valued sections"

$$D : \Omega^n \underset{A}{\otimes} \mathcal{E} \to \Omega^{n+1} \underset{A}{\otimes} \mathcal{E}, \quad D(\omega \underset{A}{\otimes} s) = \mathrm{d}\omega \underset{A}{\otimes} s + (-1)^n \omega Ds, \quad \forall \omega \in \Omega^n.$$

We can then define the curvature $F_D = D^2 : \mathcal{E} \to \Omega^2 \otimes_A \mathcal{E}$.

Proposition 4.2. *Let $\mathcal{E}$ be a vector bundle over A with projector e, so $\mathcal{E} = A^N.e$ for some number N copies of A and $e \in M_N(A)$. Then*

$$D(\tilde{v}.e) = \mathrm{d}(\tilde{v}.e).e = \mathrm{d}\tilde{v}.e + \tilde{v}.\mathrm{d}e.e, \quad \tilde{v} \in A^N$$

is a connection, the "Grassmann connection" on $\mathcal{E}$, with curvature given by

$$F_D = -\mathrm{d}e.\mathrm{d}e.e.$$

Proof. We write an element $v \in \mathcal{E}$ as the image of a row vector $\tilde{v} = (v_1, \dots, v_N)$. The first expression for D shows that it is well defined and depends only on v. The second follows by the Leibniz rule applied component-wise and $e^2 = e$. The connection rule follows from the first form as

$$D(a\tilde{v}.e) = (\mathrm{d}(a\tilde{v}.e)).e = (\mathrm{d}a.\tilde{v}.e + a\mathrm{d}(\tilde{v}.e)).e = \mathrm{d}a.\tilde{v}.e + aD(\tilde{v}.e),$$

for all $a \in A$, $\tilde{v}.e \in \mathcal{E}$. For the curvature, we compute $D^2(\tilde{v}.e)$ as

$$D(\mathrm{d}\tilde{v}.e + \tilde{v}.\mathrm{d}e.e) = (\mathrm{d}(\mathrm{d}\tilde{v}.e)).e + (\mathrm{d}(\tilde{v}.\mathrm{d}e.e)).e$$

$$= -\mathrm{d}\tilde{v}.\mathrm{d}e.e + \mathrm{d}\tilde{v}.\mathrm{d}e.e - \tilde{v}.\mathrm{d}e.\mathrm{d}e.e$$

where we treat $\mathrm{d}\tilde{v}$ and $\tilde{v}.\mathrm{d}e$ as Ω^1-valued row vectors (a form-valued section). We used the Leibniz rule and $\mathrm{d}^2 = 0$. Also, $e^2 = e$ implies $e.\mathrm{d}e.\mathrm{d}e = (\mathrm{d}e^2)\mathrm{d}e - \mathrm{d}e.e.\mathrm{d}e = \mathrm{d}e.\mathrm{d}e^2 - \mathrm{d}e.e.\mathrm{d}e = \mathrm{d}e.\mathrm{d}e.e$ so that

$$F_D(\tilde{v}.e) = -\tilde{v}.\mathrm{d}e.\mathrm{d}e.e = -\tilde{v}.e.\mathrm{d}e.\mathrm{d}e = -\tilde{v}.e.\mathrm{d}e.\mathrm{d}e.e$$

acts simply as the $M_N(\Omega^2)$-valued matrix as stated on $\mathcal{E}$. $\qquad\square$

The converse is also true in the case of the universal calculus on suitable A; in this case a finitely generated module $\mathcal{E}$ is projective if and only if it admits a connection with the universal calculus (Cuntz–Quillen theorem). As a non-trivial example we treat $\mathbb{C}P^1$ first classically (but using our algebraic methods) and a non-commutative version.

Example 4.3. $\mathbb{C}P^1$ means the set of lines in $\mathbb{C}^2$ and has a tautological bundle $\mathcal{E}$ where E_x at each $x \in \mathbb{C}P^1$ is the line itself in $\mathbb{C}^2$. Each $E_x \cong \mathbb{C}$ and we have a rank 1 vector bundle. Now, lines in $\mathbb{C}^2$ are in one-to-one correspondence with certain matrices

$$\mathbb{C}P^1 \leftrightarrow \{(e_{ij}) \in M_2(\mathbb{C}) \mid e^2 = e, \quad \mathrm{Tr}(e) = 1, \quad e_{ij}^* = e_{ji}\}.$$

Here any such Hermitian e has real eigenvalues and since $e^2 = e$ these are each 0 or 1. Since $\mathrm{Tr}(e) = 1$, exactly one eigenvalue is 1 and the corresponding eigenvector defines a one-dimensional subspace. Using the dictionary in Theorem 1.1 the complex polynomial subset of $\mathbb{C}^4$ defined by the relations corresponds to an algebra $\mathbb{C}[\mathbb{C}P^1]$ defined as the $*$-algebra with generators e_{ij}, $i, j = 1, 2$, relations $e^2 = e$, $\mathrm{Tr}e = 1$ and $*$-structure $e_{ij}^* = e_{ji}$. The section of the tautological vector bundle is the projective module $\mathcal{E} = \mathbb{C}[\mathbb{C}P^1]^2.e$ where $e \in M_2(\mathbb{C}[\mathbb{C}P^1])$ is the matrix of generators. The Grassmann connection recovers us the monopole connection in $\mathbb{C}P^1 \cong S^2$. Note that $\mathbb{C}[\mathbb{C}P^1]$ is a "complexified $\mathbb{C}P^1$" or $S_{\mathbb{C}}^2$ and is an affine variety while the $*$-algebra structure remembers the "real form" $\mathbb{C}P^1$ or S^2. One can choose more conventional coordinates by parametrising

$$e = \begin{pmatrix} 1-x & z \\ z^* & x \end{pmatrix}$$

for self-adjoint generator x and generator z. The form of e solves the trace and Hermitian conditions and the remaining projector relations become

$$zx = xz, \quad zz^* = z^*z = x(1-x)$$

which indeed describes a sphere of radius $1/2$ if we write $z = x_1 + \imath x_2$ and $x = x_3 + \frac{1}{2}$. For the Grassmann connection note that $\mathrm{d}z.z^* + z\mathrm{d}z^* = (1-2x)\mathrm{d}x$ from d applied to the quadratic relation. Hence in one patch

$$\mathrm{de}.\mathrm{de} = \begin{pmatrix} -\mathrm{d}x & \mathrm{d}z \\ \mathrm{d}z^* & \mathrm{d}x \end{pmatrix}^2 = \begin{pmatrix} \mathrm{d}z\mathrm{d}z^* & -2\mathrm{d}x\mathrm{d}z \\ 2\mathrm{d}x\mathrm{d}z^* & -\mathrm{d}z\mathrm{d}z^* \end{pmatrix}$$

$$= \frac{\mathrm{d}z\mathrm{d}z^*}{1-2x} \begin{pmatrix} 1-2x & 2z \\ 2z^* & -(1-2x) \end{pmatrix} = \frac{\mathrm{d}z\mathrm{d}z^*}{1-2x}(2e-1).$$

From which

$$F_D = -\frac{dz dz^*}{1 - 2x} e = -\imath \frac{dx_1 dx_2}{x_3} e$$

which is a constant multiple of the volume form (the top form) on S^2 as it should be for the monopole.

Notice in this example that we did not impose *a priori* that the $*$-algebra was commutative, this came out of the projector relations! Hence we can discover non-commutative versions just as easily.

Example 4.4. The "fuzzy sphere" $\mathbb{C}_\lambda[\mathbb{C}P^1]$ or $\mathbb{C}_\lambda[S^2]$ is defined exactly as in Example 4.3 but with $\text{Tr}(e) = 1 + \lambda$ where $\lambda \in \mathbb{R}$. The result is a $*$-algebra

$$[x, z] = \lambda z, \quad [z, z^*] = 2\lambda \left(x - \frac{1 + \lambda}{2}\right), \quad z^* z = x(1 - x)$$

and the Grassmann connection defines the "fuzzy monopole" with respect to any differential calculus. As before, we write

$$e = \begin{pmatrix} 1 + \lambda - x & z \\ z^* & x \end{pmatrix}$$

which solves the deformed trace condition. The remaining projector relations come out as stated. The first two relations are in fact the enveloping algebra $U(\text{su}_2)$ as 'fuzzy $\mathbb{R}^3$' and the last relation sets the Casimir equal to a constant, i.e. this is the standard quantisation of S^2 as a coadjoint orbit in su_2^* as mentioned at the end of Section 2.

Example 4.5. The q-sphere $\mathbb{C}_q[\mathbb{C}P^1]$ or $\mathbb{C}_q[S^2]$ is defined exactly as in Example 4.3 but with a modified "q-trace" $\text{Tr}_q(e) = e_{11} + q^2 e_{22}$ where $q \in \mathbb{R}^*$. The result is the $*$-algebra

$$zx = q^2 xz, \quad zz^* = q^4 z^* z + q^2(1 - q^2)x, \quad z^* z = x(1 - x)$$

and the Grassmann connection is the q-monopole with respect to any calculus. Here we solve the $\text{Tr}_q(e) = 1$ condition with

$$e = \begin{pmatrix} 1 - q^2 x & z \\ z^* & x \end{pmatrix}.$$

The remaining projector relations come out as stated. The origin of the q-trace is that this is $\mathbb{C}_q[\text{SU}_2]$-invariant and as a result $\mathbb{C}_q[S^2]$ necessarily has a (left) coaction of $\mathbb{C}_q[\mathrm{SU}_2]$ corresponding to the classical picture. One can well ask what happens if one combines this with the previous example and asks for $\mathrm{Tr}_q(e) = 1 + \lambda$. The result is of course the "q-fuzzy sphere" with two parameters q, λ. It turns out for $q \neq 1$ to be isomorphic to a "constant time slice" (depending on λ) of the unit hyperboloid in q-Minkowski space.

For the non-commutative torus, we work with a Schwarz space version

$$\tilde{A}_\theta = \left\{ \sum a_{mn} u^m v^n \, \middle| \, (a_{mn}) \in S(\mathbb{Z}^2) \right\},$$

where the Schwarz space means functions on $\mathbb{Z}^2$ decaying faster than any power of n, m (more precisely $(|m| + |n|)^N |a_{mn}|$ is bounded on $\mathbb{Z}^2$ for all $N \in \mathbb{N}$). This is larger than the algebraic version but still workable. One defines $S(\mathbb{R})$ similarly as functions on $\mathbb{R}$ decaying faster than any power.

Example 4.6. For irrational θ, $\tilde{A}_\theta$ has finitely generated projective modules $\mathcal{E}_{p,q}$ where $p, q \in \mathbb{Z}$. As a vector space $\mathcal{E}_{p,q} = S(\mathbb{R}, \mathbb{C}^q)$ and

$$(u.s)(x) = u_0 s \left(x - \theta + \frac{p}{q} \right), \quad (v.s)(x) = e^{2\pi i x} v_0 s(x),$$

where u_0, v_0 are fixed unitary matrices on $\mathbb{C}^q$ obeying $v_0 u_0 = e^{2\pi i \frac{p}{q}} u_0 v_0$ and $s \in S(\mathbb{R}, \mathbb{C}^q) = S(R) \otimes \mathbb{C}^q$. One may verify that this is a module much as we did in constructing A_θ in Section 1. The associated "Powers projector" $e_{p,q}$ is a bit beyond our scope.

4.2. *K-theory and cyclic cohomology*

Vector bundles are one of the tools used in geometry to obtain topological invariants. From the operation $\oplus$ among vector bundles one can construct $K_0(X)$ as an abelian group made, loosely speaking, out of stable equivalence classes. Two vector bundles are stably equivalent if they are isomorphic after direct sum with some third vector bundle. More precisely, isomorphism classes of vector bundles form a semigroup under direct sum. Given any semigroup S with operation $\oplus$ (say) its associated group consists of equivalence classes of $S \times S$ where

$$(E, F) \sim (E' F') \Leftrightarrow \exists G \in S \text{ such that } E \oplus F' \oplus G = E' \oplus F \oplus G.$$

(Grothendieck's construction). One can sloppily think of $(E, 0)$ as the "positive elements" and $(0, E)$ as their adjoined negatives. In algebraic terms, we define $K_0(A)$ similarly as direct sum of isomorphism classes of finitely generated projective modules made into a group. Elements of $S(A)$ are equivalence classes $[e]$ of projectors, where

(1) Extension by zero: $\begin{pmatrix} e & 0 \\ 0 & 0 \end{pmatrix} \in M_{N+k}(A)$
(2) Conjugation: ueu^{-1} for any $u \in \mathrm{GL}_N(A)$

define the same projective module up to isomorphism. Likewise, if the extension of one projector is "conjugate" in a suitable sense to an extension of another, we consider them equivalent members of $S(A)$. In a $*$-algebra setting we require projectors to be Hermitian and the u above to be unitary. For example, over $\mathbb{C}$, one has $K_0(S^2) = \mathbb{Z} \times \mathbb{Z}$ generated by tensor products of the monopole bundle and direct sums. We have the same picture for generic q for $K_0(\mathbb{C}_q[S^2])$ with the non-trivial generator being the $[e]$ for e in Example 4.5.

More novel is another "topological invariant" which is more like homology of a space. As usual we let A be a unital algebra over k.

Definition 4.7. The cyclic cochain complex of A is
$$C_\lambda^n(A) = \{\phi : A^{\otimes(n+1)} \to k \mid \phi(a_1, \ldots, a_n, a_0) = (-1)^n \phi(a_0, a_1, \ldots, a_n)\}$$
$$(b\phi)(a_0, \ldots, a_{n+1}) = \sum_{j=0}^{n} (-1)^j \phi(a_0, \ldots, a_j a_{j+1}, \ldots, a_{n+1})$$
$$+ (-1)^{n+1} \phi(a_{n+1} a_0, a_1, \ldots, a_n).$$
where $b : C_\lambda^n \to C_\lambda^{n+1}$ obeys $b^2 = 0$. The cyclic cohomology $HC_\lambda^n(A)$ is the kernel of b on degree n modulo the image of b on degree $n-1$.

A cyclic n-cochain is called *unital* if it vanishes when any of its arguments is 1 and $n \geq 1$. When $n = 0$ the condition is taken to be empty. To bring out the "geometric meaning" we suppose that A has an exterior algebra Ω of sufficiently high top degree and that $\int : \Omega^n \to k$ is a linear map such that
$$\int d\omega = 0, \quad \int \omega \rho = (-1)^{|\omega||\rho|} \int \rho \omega$$
on forms of appropriate homogeneous degree. Such an "n-cycle" is like integration on a closed submanifold of dimension n.

Proposition 4.8. *Let A be equipped with an exterior algebra and $\int$ an n-cycle. Then*
$$\phi(a_0, a_1, \ldots, a_n) = \int a_0 da_1 \cdots da_n$$
is a unital cyclic n-cocycle.

Proof.   We first check cyclicity

$$\phi(a_1, \ldots a_n, a_0) = \int a_1 da_2 \cdots da_0 = (-1)^{n-1} \int a_1 d(da_2 \cdots da_n.a_0)$$

$$= (-1)^{n-1} \left(\int d(a_1 da_2 \cdots da_n.a_0) - \int da_1 \cdots da_n.a_0 \right)$$

$$= (-1)^n \int da_1 \cdots da_n.a_0 = (-1)^n \phi(a_0, a_1, \ldots, a_n)$$

using the graded Leibniz rule and the first property of $\int$. Next,

$$(b\phi)(a_0, \ldots, a_{n+1}) = \int a_0 a_1 da_2 \cdots da_{n+1}$$

$$+ \sum_{j=1}^{n} (-1)^j \int a_0 da_1 \cdots d(a_j a_{j+1}) \cdots da_{n+1}$$

$$+ (-1)^{n+1} \int a_{n+1} a_0 da_1 \cdots da_n$$

and we expand the summed terms by the Leibniz rule as

$$\sum_{j=1}^{j=n} (-1)^j \int a_0 da_1 \cdots da_j.a_{j+1} da_{j+2} \cdots da_{n+1}$$

$$+ \sum_{j=1}^{j=n} (-1)^j \int a_0 da_1 \cdots da_{j-1}.a_j da_{j+1} \cdots da_{n+1}.$$

Now the interior of the first sum at j cancels with the interior of the second at $j + 1$. What remains is the boundary $j = 1$ of the second sum, which cancels with the first term of $b\phi$ above, and $j = n$ of the first sum giving $(-1)^n \int a_0 da_1 \cdots da_n.a_{n+1}$. This cancels with the last term of $b\phi$ above due to the second requirement of $\int$ an n-cycle. Note that we only really need this in the form $\int a\omega = \int \omega a$ for all $a \in A$ and $\omega \in \Omega^n$ but this implies and is therefore equivalent to the graded version for general $\rho \in \Omega$ (a short proof by induction on degree of ρ). Clearly ϕ is unital since $d1 = 0$.   □

The converse is also true in the case of the universal calculus on A; in this case the two notions are equivalent. Starting with a unital n-cocycle ϕ we define $\int$ on degree n by the same formula as above but read the other way. In degree 0 an element of $HC^0(A)$ and a 0-cycle both mean a "trace", i.e. a map $\phi : A \to k$ such that $\phi(ab) = \phi(ba)$. In degree 1 a unital 2-cocycle is a linear map $\phi : A^{\otimes 2} \to k$ with

$$\phi(1, a) = 0, \quad \phi(a_0 a_1, a_2) - \phi(a_0, a_1 a_2) + \phi(a_2 a_0, a_1) = 0$$

(which implies that ϕ is antisymmetric). Now $\int a_0 da_1 = \phi(a_0, a_1)$ is well-defined because if $\sum a_0^\alpha da_1^\alpha = 0$ then $\sum a_0^\alpha \otimes a_1^\alpha = \sum a_0^\alpha a_1^\alpha \otimes 1$. Writing $a_1^\alpha = \lambda^\alpha + b^\alpha \in k1 \oplus A'$ (choosing a complement of $k1$) we find $\sum a_0^\alpha \otimes b^\alpha = \sum a_0^\alpha b^\alpha \otimes 1$ and conclude that $\sum a_0^\alpha \otimes b^\alpha = 0$ and hence $\sum a_0^\alpha \otimes a_1^\alpha \in A \otimes 1$. Hence in this case $\sum \phi(a_0^\alpha, a_1^\alpha) = 0$ as required. Once we know that $\int$ is well defined, we just push the proof of the proposition in reverse; by definition $\int da = \phi(1, a) = 0$ while the cocycle condition amounts $\int a_0 da_1 . a_2 = \int a_2 a_0 da_1$ as required.

Example 4.9. $\int \sum a_{mn} u^m v^n = a_{00}$ defines a 0-cycle on the non-commutative torus $\mathcal{A}_\theta$. Remembering that the a_{mn} are at $\theta = 0$ the Fourier coefficients of a function on $S^1 \times S^1$, this becomes the Haar integral on the torus. For the calculus in Example 3.6 we define a 2-cycle by $\int a e_1 e_2 = \int a$ for $a \in \mathcal{A}_\theta$. This defines a unital 2-cocycle on $\mathcal{A}_\theta$. To see what this looks like, we use as basic 1-forms $e_1 = u^{-1} du$ and $e_2 = v^{-1} dv$ which commute with $\mathcal{A}_\theta$ and anticommute as classically among themselves. In that case $da = (\partial_u a) e_1 + (\partial_v a) e_2$ defines two derivations $\partial_u, \partial_v : \mathcal{A}_\theta \to \mathcal{A}_\theta$. They look as classically $\partial_u = u^{-1} \frac{\partial}{\partial u}$ and similarly for ∂_v provided we understand all expressions as "normally ordered" with u to the left of v. We have

$$\phi(a, b, c) = \int a(\partial_u b e_1 + \partial_v b e_2)(\partial_u c e_1 + \partial_v c e_2) = \int a(\partial_u b \partial_v c - \partial_v b \partial_u c) e_1 e_2$$

$$= \int a(\partial_u b \partial_v c - \partial_v b \partial_u c).$$

Finally, we can put together these ideas.

Theorem 4.10. *Suppose that k has characteristic* 0. *There is a "Chern–Connes" pairing between $K_0(A)$ and $HC^{2m}(A)$ given by*

$$\langle [e], [\phi] \rangle = \frac{1}{m!} \sum_{i_0, \ldots, i_{2m}} \phi(e_{i_0 i_1}, e_{i_1 e_2}, \ldots, e_{i_{2m} i_0}) = \frac{1}{m!} \mathrm{Tr} \phi(e, e, \ldots, e)$$

Proof. We sketch why this well defined on both sides. If we change e by extension by zero then clearly the right-hand side does not change. The right-hand side is also unchanged if we conjugate $e \in M_N(A)$. In degree 0 this is $\phi(u_{ij} e_{jk} u_{ki}^{-1}) = \phi(u_{ki}^{-1} u_{ij} e_{jk}) = \phi(e_{ii})$ (summations understood) by the cocycle (trace) requirement. For higher degree we similarly need to use the cocycle condition. For example, for degree 2 this is

$$\phi(a_0 a_1, a_2, a_3) - \phi(a_0, a_1 a_2, a_3) + \phi(a_0, a_1, a_2 a_3) - \phi(a_3 a_0, a_1, a_2) = 0.$$

Suppose for the moment that $N = 1$ so $e, u \in A$. Then using $e^2 = e$,

$$\phi(ueu^{-1}, ueu^{-1}, ueu^{-1}) = \phi(ue.eu^{-1}, ueu^{-1}, ueu^{-1})$$
$$= \phi(ue, eu^{-1}, ueu^{-1})$$

where we applied the cocycle condition with $a_0 = ue, a_1 = eu^{-1}, a_2 = a_3 = ueu^{-1}$. Similarly, the right-hand side is equal to $\phi(ue, eu^{-1}, ue.eu^{-1}) = 2\phi(ue, e, eu^{-1}) - \phi(ue, eu^{-1}, ue.eu^{-1})$, where we use that ϕ is invariant under cyclic rotations. Hence, as 2 is invertible, we obtain $\phi(ue, e, eu^{-1}) = \phi(ue, e.e, eu^{-1}) = 2\phi(ue, e, eu^{-1}) - \phi(e, e, e)$. Hence we obtain $\phi(e, e, e)$. Now when $N > 1$ we have matrix indices on the u, e being matrix multiplied. As Tr on a product of matrices is invariant under cyclic rotations used in some of the above steps, the same steps still work. The proof for general even degree is best done by more sophisticated arguments.

On the other side, if $\phi = b\psi$ we have

$$\mathrm{Tr}\left(\sum_{j=0}^{2m-1} (-1)^j \psi(e, e, \ldots, e) + (-1)^{2m} \psi(e, e, \ldots, e) \right) = \mathrm{Tr}\, \psi(e, e, \ldots, e)$$

since $e^2 = e$ and there are an odd number of alternating such terms. But ψ is odd and cyclic, hence changes sign under cyclic rotation of its arguments while the trace does not change under the cyclic rotation of the corresponding matrix products. Hence $\mathrm{Tr}\, \psi(e, e, \ldots, e) = 0$. $\qquad\square$

The interpretation of this is best seen if ϕ is given by a $2m$-cycle. Then

$$\langle [e], [\phi] \rangle = \frac{1}{m!} \mathrm{Tr} \int ede \cdots de = \frac{(-1)^m}{m!} \mathrm{Tr} \int F_D^m = \mathrm{Tr} \int e^{-F_D}$$

for D the Grassmann connection. Here $F_D = -edede = -dede.e$ as we noted before, so we may move all the e's to the left. Also, as in differential geometry, we consider $\int$ to be zero on forms of the wrong degree. Here e^{-F_D} it is the analogue of the classical Chern character ch : $K_0(X) \to H_{DR}(X)$. The classical de Rham cohomology pairs with homology in X represented by cycles $C \subseteq X$ and the pairing is $\int_C$ on forms of correct degree.

Example 4.11. We consider $\mathbb{C}_q[S^2]$ with the tautological or monopole bundle in Example 4.5. For $q \neq 1$ this has a certain 0-cocycle ϕ defined by

$$\phi(z^m) = \phi(z^{*m}) = 0, \quad \phi(z^m x^n) = \phi(z^{*m} x^n) = \frac{\delta_{m,0}}{1 - q^{2n}}, \quad n > 0.$$

Then

$$\langle [e], [\phi] \rangle = \phi \left(\mathrm{Tr} \begin{pmatrix} 1 - q^2 x & z \\ z^* & x \end{pmatrix} \right) = \phi(1 + (1 - q^2)x) = 1.$$

This shows that the q-monopole bundle is non-trivial. The 1 here is the topological charge of the q-monopole. We use basis $\{z^m x^n \mid m > 0, n \geq 0\} \cup \{z^{*m} x^n \mid m > 0, n \geq 0\} \cup \{x^n \mid n \geq 0\}$ for $\mathbb{C}_q[S^2]$ in view of its relations. Note that ϕ has no limit as $q \to 1$ and although broadly similar to it, is *not* the restriction of the Haar integral on $\mathbb{C}_q[SU_2]$ discussed in Section 2. The latter obeys $\int ba = \int \sigma(a)b$ for all a, b, where σ is a certain "twisting automorphism". Thus we have to use something different to fit the standard axioms of cyclic cohomology *or* we can use $\int$ but have to use a q-deformed or "twisted" cyclic cohomology to properly accommodate such q-deformed geometries in a way that has a classical limit as $q \to 1$.

5. Non-commutative Riemannian Geometry

We now take a deeper point of view on both vector bundles and Riemannian geometry, starting with the notion of principal G-bundle P over a manifold X. This is defined exactly like a vector bundle with a surjection $\pi : P \to X$ but each fibre $P_x = \pi^{-1}(x)$ now has the structure of a fixed group G. This is achieved smoothly by starting with a free right action of G on the manifold P such that $X = P/G$. A connection on P is defined concretely as an equivariant complement in $\Omega^1(P)$ to the "horizontal forms" (those pulled back from $\Omega^1(X)$). This is, however, equivalent to $\omega \in \Omega^1(P) \otimes \mathfrak{g}$ with certain properties. Here $\mathfrak{g}$ is the Lie algebra of G. Given this data, there is an associated vector bundle $E = P \times_G V$ and a connection D on it, for any representation V of G.

As an example, the frame bundle on a manifold of dimension n is a certain principal SO_n-bundle. In non-commutative geometry we would not know which quantum group version of this to take or even what n would be if there was no cotangent dimension, so we are forced to generalise the notion even in classical differential geometry to that of a *G-framing*. This means a general principal G-bundle P, a representation

V and $\theta \in \Omega^1(P, V^*)$ that is G-equivariant and horizontal such that the induced map $\mathcal{E} \to \Omega^1(X)$ on the associated bundle sections given by multiplication by θ pointwise (by contracting the V^* of θ with the V-value of a section) is an isomorphism. The entries of θ in local coordinates are the "n-bein". In this context, the framing isomorphism turns the covariant derivative D induced by a "spin connection" ω on the principal bundle into a connection ∇ on Ω^1. This being torsion free amounts to $D\theta = 0$ when θ is itself viewed as a section of an associated bundle. In classical geometry we would also want to choose ω such that ∇ is metric compatible, hence the Levi-Civita connection for the metric. However, for a general G-framing, and also in the quantum case, this may not be possible and we are forced to a weaker concept of Riemannian geometry. One such is the notion of both a G-framing and a G-coframing. Given the former, the latter just means θ^* so that (V^*, θ^*) is also a framing, which is equivalent to Ω^1 now being isomorphic to $\mathcal{E}^*$, i.e. to Ω^1 being isomorphic to its own dual, i.e. to a possibly non-symmetric metric $g = \langle \theta^*, \theta \rangle \in \Omega^1 \otimes_{C^\infty(X)} \Omega^1$ where the angular brackets denote evaluation of V^*, V and the result lies in the tensor square of $\Omega^1(X)$ due to equivariance and horizontality of θ, θ^*. In this context we can ask that ω is torsion free with respect to θ^*, i.e. $D\theta^* = 0$. We call this condition 'cotorsion free' and the remarkable thing is that if the connection is already torsion free then cotorsion free amounts to $(\wedge \otimes \mathrm{id})\nabla g = 0$, a weaker classical notion of metric compatibility but which tends to work in the non-commutative case and which one can solve for directly.

5.1. *Quantum principal bundles and framing*

We now turn to the non-commutative formulation on an algebra A with structure quantum group H.

Definition 5.1. A "quantum principal bundle" over A means:

(1) P a right H-comodule algebra via $\Delta_R : P \to P \otimes H$ with $A = P^H$.
(2) Compatible differential structures with $\Omega^1(H) = H.\Lambda^1_H$ bicovariant, $\Omega^1(P)$ right H-covariant and $\Omega^1(A) = A(\mathrm{d}A)A \subseteq \Omega^1(P)$.
(3) $0 \to P\Omega^1(A)P \to \Omega^1(P) \to P \otimes \Lambda^1_H \to 0$ is exact.

Here the map ver $: \Omega^1(P) \to P \otimes \Lambda^1_H$ is the analogue of the "vertical vector fields" generated by the Lie algebra of the fibre group acting on the total space of the bundle. Here $\Lambda^1_H = H^+/I_H$ is like the dual of the

Lie algebra so ver is like a coaction version of the vertical vector fields. It is defined at the level of universal calculi ver : $\Omega^1_{\text{univ}}(P) \to P \otimes H^+$ by $\text{ver}(u \otimes v) = u\Delta_R v$. Compatibility of the calculi means

$$\Delta_R N_P \subseteq N_P \otimes H, \quad \text{ver}(N_P) \subseteq P \otimes I_H, \quad N_A = N_P \cap \Omega^1_{\text{univ}}(A)$$

where we recall that calculi on algebras are defined by subbimodules N, and ensures that ver descends to the non-universal calculi. If we already have a bundle with the universal calculi then the further content of (3) is that we have in fact $\text{ver}(N_P) = P \otimes I_H$.

Surjectivity of the condition (3) corresponds classically to freeness of the action (i.e. only the identity element fixes all points), while exactness in the middle says that the horizontal forms $P\Omega^1(A)P$ are exactly those killed by the "vertical vector fields". This algebraic and global condition in non-commutative geometry replaces the concept of a local trivialisation, which would normally be used to prove such things in classical differential geometry. Following the classical geometry, a connection on a quantum bundle means a left P-module and right H-comodule map

$$\Pi : \Omega^1(P) \to \Omega^1(P), \quad \Pi^2 = \Pi, \quad \ker \Pi = P\Omega^1(A)P$$

i.e. an equivariant complement to the horizontal forms. A connection is called *strong* if $(\text{id} - \Pi)\mathrm{d}P \subseteq \Omega^1(A)P$, which is automatic if $\Omega^1(A)$ commutes with all elements of P. We recall from Section 3.2 that H coacts on Λ^1_H by Ad_R in the bicovariant case.

Proposition 5.2. *Connections on a quantum principal bundle P correspond to right comodule maps $\omega : \Lambda^1_H \to \Omega^1(P)$ such that $\text{ver} \circ \omega = 1 \otimes \text{id}$.*

Proof. One can check that ver is equivariant where $P \otimes \Lambda^1_H$ has the tensor product of Δ_R on P and Ad_R in Corollary 3.10 on Λ^1_H. Then given ω, set

$$\Pi = \cdot (\text{id} \otimes \omega) \circ \text{ver}$$

which is then equivariant, and clearly a projection due to $\text{ver} \circ \omega = 1 \otimes \text{id}$. Conversely, given Π, and exactness of our sequence in Definition 5.1 we define

$$\omega(v) = \Pi \circ \text{ver}^{-1}(1 \otimes v), \quad \forall v \in \Lambda^1_H,$$

meaning we choose any element mapping onto $1 \otimes v$ under ver, then apply Π. This is well defined just because Π and ver have the same kernel. $\qquad\square$

Non-trivial examples are provided by quantum homogeneous spaces. Here the "total space" algebra P is itself a quantum group with left-covariant calculus and there is a Hopf algebra surjection $\pi : P \to H$ (so in the classical case a subgroup). There is then a canonical coaction $\Delta_R = (\mathrm{id} \otimes \pi)\Delta$ and we suppose that this gives a quantum bundle with the universal calculus. The condition

$$(\mathrm{id} \otimes \pi)\mathrm{Ad}_R(I_P) \subseteq I_P \otimes H, \quad \pi(I_P) = I_H$$

then gives a quantum bundle over $A = P^H$ with our given calculi. We also have a left coaction $\Delta_L = (\pi \otimes \mathrm{id})\Delta$ and if $i : H \to P$ is a linear map with

$$\pi \circ i = \mathrm{id}, \quad \Delta_R \circ i = (i \otimes \mathrm{id})\Delta, \quad \Delta_L \circ i = (\mathrm{id} \otimes i)\Delta, \quad i(I_H) \subseteq I_P$$

then $\omega(h) = \sum Si(h)_{(1)}di(h)_{(2)}$ for any $h \in H^+$ is a strong connection.

Example 5.3. Over $\mathbb{C}$, let $P = \mathbb{C}_q[\mathrm{SL}_2]$ with its 3D left-covariant calculus and $H = \mathbb{C}_{q^2}[t, t^{-1}]$ with calculus $dt.t = q^2 t dt$, as in Examples 3.12 and 3.14. Let

$$\pi : \mathbb{C}_q[\mathrm{SL}_2] \to \mathbb{C}[t, t^{-1}], \quad \pi \begin{pmatrix} a & b \\ c & d \end{pmatrix} = \begin{pmatrix} t & 0 \\ 0 & t^{-1} \end{pmatrix}$$

which induces a coaction $\Delta_R f = f \otimes t^{|f|}$ where $|f|$ is the degree on a monomial f as in Example 3.14. This is clear from

$$\Delta_R \begin{pmatrix} a & b \\ c & d \end{pmatrix} = \begin{pmatrix} a \otimes t & b \otimes t^{-1} \\ c \otimes t & d \otimes t^{-1} \end{pmatrix}.$$

The coinvariants subalgebra $P^H = \mathbb{C}_q[S^2_{\mathbb{C}}]$ therefore means the degree 0 elements, generated by $x = -q^{-1}bc$, $z = cd$ and $w = -qab$. We identify them with $\mathbb{C}_q[S^2_{\mathbb{C}}]$ in Example 4.5 with $wx = q^{-2}xw$ for the relations of $w = z^*$ there. The calculus on $\mathbb{C}_q[\mathrm{SL}_2]$ corresponds to

$$I_P = \langle a + q^2 d - (1 + q^2), b^2, c^2, bc, (a - 1)b, (d - 1)c \rangle$$

and one may verify that $\pi(I_P) = I_H = \langle (t - 1)(t - q^2) \rangle$, as well as the other technical conditions for a quantum bundle. The map

$$i(t^n) = a^n, \quad i(t^{-n}) = d^n, \quad \forall n \geq 0$$

has the desired bicovariance properties and gives

$$\omega(t^n - 1) = [n]_{q^2} e_0, \quad F_\omega(t^n - 1) = q^3 [n]_{q^2} e_+ \wedge e_-.$$

For brevity we take $\omega(1) = 0$. Then $\omega(t) = (Sa_{(1)})da_{(2)} = dda - qbdc = e_0$ in the 3D calculus, so the claim holds for $n = 1$. When $n \geq 2$,

$$\omega(t^n) = (Sa^n{}_{(1)}da^n{}_{(2)}) = S(a_{(1)}a'{}_{(1)}a''{}_{(1)} \cdots)d(a_{(2)}a'{}_{(2)}a''{}_{(2)} \cdots)$$
$$= S(a'{}_{(1)}a''{}_{(1)} \cdots)Sa_{(1)}((da_{(2)})a'{}_{(2)}a''{}_{(2)} \cdots + a_{(2)}d(a'{}_{(2)}a''{}_{(2)} \cdots))$$
$$= \omega(t^{n-1}) + S(a'{}_{(1)}a''{}_{(1)} \cdots)\omega(t)a'{}_{(2)}a''{}_{(2)} \cdots$$
$$= \omega(t^{n-1}) + S(a'{}_{(1)}a''{}_{(1)} \cdots)e_0 a'{}_{(2)}a''{}_{(2)} \cdots = \omega(t^{n-1}) + q^{2(n-1)}e_0$$

where $a^n = aa'a'' \cdots$ is the product of n copies of the generator $a \in \mathbb{C}_q[SL_2]$ (the primes are to keep the instances apart). We used the antimultiplicativity of the antipode S and the Leibniz rule for the third equality. For the last equality we used that $a'{}_{(2)}a''{}_{(2)} \cdots$ has degree $n - 1$ and hence its commutation relations with e_0 give a factor $q^{2(n-1)}$ after which we cancel using the antipode axioms and $\epsilon(a) = 1$. The computation for negative n is similar and the curvature computation is an exercise.

To complete the theory, let V be a finite-dimensional right H-comodule. We have an "associated bundle" $\mathcal{E} = (P \otimes V)^H$ which is now a vector bundle as in Section 4.1 at least if there is a connection at the universal calculus level (so that it is projective). A strong connection ω, under some technical assumptions on the bundle, defines

$$D : \mathcal{E} \to \Omega^1(A) \underset{A}{\otimes} \mathcal{E}, \quad D = (\mathrm{id} - \Pi)d$$

as the associated covariant derivative or vector bundle connection on $\mathcal{E}$.

Example 5.4. For the quantum principal bundle in Example 5.3, let $V = \mathbb{C}.v$ be one-dimensional and $\Delta_R v = v \otimes t^n$. The associated bundle $\mathcal{E}_n = \mathbb{C}_q[SL_2]_{-n}$ (the degree $-n$ part) defines the q-monopole of charge n. If $u \in \mathcal{E}_n$ then $\mathrm{ver}(du) = u^{(\bar{1})} \otimes u^{(\bar{2})} - u \otimes 1 = u \otimes (t^{-n} - 1)$ projected to Λ^1_H. Hence $\Pi(du) = u\omega(t^{-n} - 1)$ and

$$Du = (\mathrm{id} - \Pi)du = du - u\omega(t^{-n} - 1) = du + uq^{-2n}[n]_{q^2}e_0$$

for the q-monopole connection in Example 5.3.

Next we define a framing of A as a quantum principal bundle and $\theta : V \to P\Omega^1(A)$ such that the induced left A-module map

$$s_\theta : \mathcal{E} \to \Omega^1(A), \quad p \otimes v \mapsto p\theta(v)$$

is an isomorphism.

Theorem 5.5. *Let* $\pi : P \to H$ *be a quantum homogeneous bundle as above. Then* $A = P^H$ *is framed by the bundle and*

$$V = (P^+ \cap A)/(I_P \cap A), \quad \Delta_R v = \tilde{v}_{(2)} \otimes S\pi(\tilde{v}_{(1)}), \quad \theta(v) = S\tilde{v}_{(1)}\mathrm{d}\tilde{v}_{(2)}$$

where $\tilde{v}$ *is a representative of* v *in* $P^+ \cap A$. *Hence every quantum homogeneous space is a "quantum manifold" in the framed sense.*

Proof. First observe that $v \in A$ means $v_{(1)} \otimes \pi(v_{(2)}) = v \otimes 1$. Moreover, if $v \in A$ then $v_{(1)} \otimes v_{(2)} \in P \otimes A$ because $v_{(1)} \otimes v_{(2)(1)} \otimes \pi(v_{(2)(2)}) = v_{(1)(1)} \otimes v_{(1)(2)} \otimes \pi(v_{(2)}) = v_{(1)} \otimes v_{(2)} \otimes 1$, and if $v \in P^+ \cap A$ then $\epsilon(v_{(2)})\pi(Sv_{(1)}) = \pi(Sv) = S\pi(v_{(2)})\epsilon(v_{(1)}) = 1\epsilon(v) = 0$ so that Δ_R restricts to $P^+ \cap A$. Similarly,

$$\Delta_R v = v_{(1)} \otimes \pi(Sv_{(1)}) = v_{(1)(2)} \otimes \pi(Sv_{(1)(1)})\pi(v_{(2)}) = v_{(2)} \otimes \pi(Sv_{(1)}v_{(3)})$$

which is the projected adjoint coaction. I_P is stable under this, hence if $v \in I_P \cap A$ then $\Delta_R v \in I_P \cap A \otimes H$ and Δ_R descends to V. Meanwhile, if $v \in I_P$ then $S\tilde{v}_{(1)} \otimes \tilde{v}_{(2)} \in N_P$ and hence $\theta(v) = 0$ in $\Omega^1(P)$, so this is well-defined. Moreover, if $\tilde{v} \in A$ is a representative of $v \in V$ then $\theta(v) = S\tilde{v}_{(1)}\mathrm{d}\tilde{v}_{(2)} \in P\Omega^1(A)$ as required. Hence all maps are defined as required and we have $s_\theta : (P \otimes V)^H \to \Omega^1(A)$. We provide its inverse by quotienting the inverse in the universal calculus case, namely

$$s_\theta^{-1}(a\mathrm{d}b) = [ab_{(1)} \otimes b_{(2)} - ab \otimes 1], \quad \forall a, b \in A$$

where the expression in square brackets lies in $P \otimes P^+ \cap A$ and $[\,]$ denotes the equivalence class modulo $I_P \cap A$. That the result actually lies in $(P \otimes V)^H$ and gives the inverse of s_θ is then a direct verification. $\square$

Example 5.6 ([5]). At least for generic q, $A = \mathbb{C}_q[S_{\mathbb{C}}^2] = \mathbb{C}_q[\mathrm{SL}_2]_0$ in Examples 4.5 and 5.3 is framed, with

$$\Omega^1 \cong \mathcal{E}_{-2} \oplus \mathcal{E}_2.$$

We identify the summands with the holomorphic and antiholomorphic quantum differentials $\Omega^{1,0}$, $\Omega^{0,1}$ in a double complex. There is also a coframing giving us the quantum metric on $\mathbb{C}_q[S_{\mathbb{C}}^2]$ as

$$g = q\mathrm{d}w \underset{A}{\otimes} \mathrm{d}z + q^{-1}\mathrm{d}z \underset{A}{\otimes} \mathrm{d}w + q[2]_{q^2}\mathrm{d}x \underset{A}{\otimes} \mathrm{d}x$$

and the q-monopole connection induces

$$\nabla \mathrm{d}z = -q^{-1}[2]_{q^2}zg, \quad \nabla \mathrm{d}w = -q^{-1}[2]_{q^2}wg,$$

$$\nabla \mathrm{d}x = q^{-1}[2]_{q^2}\left(x - \frac{1}{[2]_{q^2}}\right)g.$$

which is torsion free, cotorsion free and q-deforms the classical Levi-Civita connection on the sphere. Here $w = z^*$ here for the real form with q real.

5.2. Bimodule connections

The above theory applies much more widely than just to q-deformations. For example, we can take A the Hopf algebra of function on a finite group with differential structure as in Example 3.13, $P = A \otimes H$ with H a second copy, $\Delta_R = \mathrm{id} \otimes \Delta$ and the canonical quantum metric $g = \sum_a e_a \otimes_A e_a$, say, and then solve for torsion-free, cotorsion-free connections. For example, the permutation group S_3 with the $\Omega(S_3)$ mentioned previously gets a non-commutative Riemannian geometry of constant curvature. In fact, the previous q-sphere and many other examples can be characterised entirely at the level of $\Omega^1(A)$ and a metric $g \in \Omega^1 \otimes_A \Omega^1$ even without a full framing picture. Here torsion, cotorsion and the Riemann curvature appear as

$$T_\nabla := \wedge \nabla - \mathrm{d} : \Omega^1 \to \Omega^2,$$

$$coT_\nabla := (\mathrm{d} \otimes \mathrm{id} - (\wedge \otimes \mathrm{id})(\mathrm{id} \otimes \nabla))g \in \Omega^2 \underset{A}{\otimes} \Omega^1,$$

$$R_\nabla := (\mathrm{d} \otimes \mathrm{id} - (\wedge \otimes \mathrm{id})(\mathrm{id} \otimes \nabla))\nabla : \Omega^1 \to \Omega^2 \underset{A}{\otimes} \Omega^1.$$

The property of being a connection is something we already covered for vector bundles in Section 4.1 as

$$\nabla(a\omega) = a\nabla\omega + \mathrm{d}a \underset{A}{\otimes} \nabla\omega, \quad \forall a \in A, \ \omega \in \Omega^1.$$

Non-degeneracy of the metric, meanwhile, is existence of a bimodule map $(\ ,\ ) : \Omega^1 \otimes_A \Omega^1 \to A$ such that

$$(\mathrm{id} \otimes (\ ,\omega))g = \omega = ((\omega,\ ) \otimes \mathrm{id})g, \quad \forall \omega \in \Omega^1.$$

It turns out that the existence of $(\ ,\ )$ forces the metric g to be central, and we also typically require "quantum symmetry" in the form $g \in \ker \wedge$ so as to have a symmetric tensor in the classical limit. This too holds for $\mathbb{C}_q[S^2]$ and $\mathbb{C}(S_3)$. In fact, the connections in these and many other examples are *bimodule connections*, meaning there exists a bimodule map $\sigma : \Omega^1 \otimes_A \Omega^1 \to \Omega^1 \otimes_A \Omega^1$ with

$$\nabla(\omega a) = (\nabla\omega)a + \sigma(\omega \otimes \mathrm{d}a), \quad \forall a \in A, \ \omega \in \Omega^1.$$

If σ exists then it is uniquely determined, so this is a condition on the connection ∇, not additional data. In this case the connection extends naturally to tensor products, so full metric compatibility makes sense as

$$\nabla g := (\nabla \otimes \mathrm{id} + (\sigma \otimes \mathrm{id})(\mathrm{id} \otimes \nabla))g = 0$$

although not the case for the mentioned examples which are merely cotorsion free. The above then amounts to a self-contained formulation of "noncommutative Riemannian geometry" at the level an algebra A equipped with a differential structure and data g, ∇ defined relative to it. This misses the deeper structure needed, say, for the Dirac operator, but has the merit that you can take your favourite algebra A and directly solve for its moduli of quantum Riemannian geometries. In the $*$-algebra case we have a notion of "real" metric (namely invariant under $(* \otimes *)$flip) and "real" connection.

Example 5.7 ([6]). We take the $*$-algebra A with generators x, t and relations $[x, t] = \lambda x$ where $x^* = x, t^* = t$ and λ is an imaginary parameter. This is a Hopf algebra (the enveloping algebra of a non-Abelian Lie algebra) with additive Δ on the generators and its natural bicovariant calculus

$$[\mathrm{d}x, x] = [\mathrm{d}x, t] = 0, \quad [x, \mathrm{d}t] = \lambda \mathrm{d}x, \quad [t, \mathrm{d}t] = \lambda \mathrm{d}t$$

admits a unique 1-parameter form of "real" quantum-symmetric metric up to normalisation, namely a λ-deformation of

$$g = b^2 x^2 \mathrm{d}t \underset{A}{\otimes} \mathrm{d}t + (1 + bt^2)\mathrm{d}x \underset{A}{\otimes} \mathrm{d}x - bxt(\mathrm{d}x \underset{A}{\otimes} \mathrm{d}t + \mathrm{d}t \underset{A}{\otimes} \mathrm{d}x)$$

where b is a non-zero real parameter. This in turn admits a unique "real" torsion-free metric compatible bimodule connection having a classical limit as $\lambda \to 0$ (there is also another "quantum Levi-Civita connection" which blows up as $\lambda \to 0$). The geometry forced out of the differential algebra is that of an expanding "big bang" type universe if $b > 0$ and a very strong gravitational source at $x = 0$ if $b < 0$, so strong that all time-like geodesics curve back in. This is not the only calculus; another choice forces one canonically to a λ-deformed Bertotti–Robinson metric and a quantum Levi-Civita bimodule connection for it.

Non-commutative geometry clearly has potential across mathematics. There is also potential to explore the structure of actual quantum systems, for example in quantum information. Better explored to date in mathematical physics is its use in modelling quantum gravity corrections to spacetime, of which the above is an example. Here it is well known that a measuring device cannot resolve distances smaller than the wavelength used. However, particles of smaller wavelength also have higher energy in quantum theory. Hence to probe smaller and smaller distances you will need heavier and heavier particles until you reach the point where the probe particles are so heavy that they curl up the geometry that you are trying to probe (form black holes in the extreme case). This means that distances less than the *Planck scale* $|\lambda| = 10^{-33}$cm make no sense and continuum differential geometry does not apply. Instead, quantum fuzziness expressed in non-commutative spacetime may appear at this scale.

6. Exercises

(1) Prove that a commutative C^*-algebra is reduced.

(2) Prove that the antipode of a Hopf algebra A is unique, antimultiplicative and anticomultiplicative in the sense $\Delta S = (S \otimes S)\mathrm{flip}\Delta$.

(3) Let u, v generate S_3 with $u^2 = v^2 = e$ and $uvu = vuv$ (denoted w). Show that $\Omega(S_3)$ defined by $\{u, v, w\}$ in Example 3.13 has dimensions 1:3:4:3:1.

(4) Show that D for the charge 1 q-monopole in Example 5.4 is isomorphic to the Grassmann connection in Proposition 4.2 for the projector in Example 4.5.

(5) In $\Omega(S_3)$ with $g = \sum_a e_a \otimes_A e_a$, show that $\nabla e_u = -e_u \otimes_A e_u - e_v \otimes_A e_w - e_w \otimes_A e_v + \frac{1}{3}\theta \otimes_A \theta$ is torsion free and cotorsion free. Here $\theta = \sum_a e_a$.

Solutions

(1) If x in the C^*-algebra is nilpotent then x^*x is also nilpotent and self-adjoint. So it is enough to show that there are no self-adjoint nilpotents. If y is self-adjoint with $y^2 = 0$ then $0 = ||y^2|| = ||y^*y|| = ||y||^2$ and hence $y = 0$. If x is self-adjoint and $x^n = 0$, $x^{n-1} \neq 0$ and $n > 2$, let $y = x^{n-1}$. Then y is self-adjoint and $y^2 = 0$ hence $y = 0$, which is a contradiction.

(2) If S' is another antipode then $S'a = \epsilon(a_{(1)})S'a_{(2)} = (Sa_{(1)(1)})a_{(1)(2)}S'a_{(2)} = (Sa_{(1)})a_{(2)(1)}(S'a_{(2)(2)}) = (Sa_{(1)})\epsilon(a_{(2)}) = Sa$. For $S(ab)$ start by proving $(S(a_{(1)}b_{(1)}))a_{(2)}b_{(2)} \otimes b_{(3)} \otimes a_{(3)} = 1 \otimes b \otimes a$, then apply S in the middle and multiply the first two factors to obtain $(S(a_{(1)}b))a_{(2)} \otimes a_{(3)} = Sb \otimes a$. Now apply S to the second factor and multiply. Deduce the last part by arrow-reversal of the preceding part.

(3) We let $e_u = \omega(\delta_u)$, etc., then $\Psi(e_u \otimes e_u) = e_u \otimes e_u$, $\Psi(e_u \otimes e_v) = e_w \otimes e_u$ etc. It follows that $e_u^2 = 0$ etc., and $e_u e_v + e_v e_w + e_w e_u = 0$,

$e_v e_u + e_u e_w + e_w e_v = 0$ as these elements are Ψ-invariant in the tensor square so vanish under the wedge product, and are a basis of such elements. Hence Λ^2 is $9 - 5 = 4$-dimensional. Using these relations one then finds three independent 3-forms in Λ^3 and one top-form $e_u e_v e_u e_w$, say. This is a finite-set calculus as in Example 3.3 for the Cayley graph of the generators.

(4) We have $Dd = dd + q^{-1}de_0 = ce_-$ and similarly $Db = ae_-$ using the formula in Example 5.4 and the relations of the 3D calculus in Example 3.14. Also from these we compute the differentials on $A = \mathbb{C}_q[S_{\mathbb{C}}^2]$ as

$$\mathrm{d}z = c^2 e_- + d^2 e_+, \quad \mathrm{d}w = -q(a^2 e_- + b^2 e_+), \quad \mathrm{d}x = -ace_- - bde_+$$

and solve for Dd, Db in terms of these. Also factorise the projector e, so

$$e = \begin{pmatrix} d \\ -b \end{pmatrix} (a \; q^{-1}c), \quad D\begin{pmatrix} d \\ -b \end{pmatrix} = \mathrm{d}e \otimes \begin{pmatrix} d \\ -b \end{pmatrix}$$

from which it is clear that $u = fd - gb \mapsto (f, g)e = (ua, uq^{-1}c)$ for $f, g \in A$ is the required isomorphism $\mathcal{E}_1 \to A^2 e$.

(5) Here $\wedge \nabla e_u = -e_v e_w - e_w e_v = \mathrm{d}e_u$ etc., by the Maurer–Cartan equations. For the cotorsion we apply wedge to the first two factors of $(\mathrm{id} \otimes \nabla)(g)$ and compare with $(\mathrm{d} \otimes \mathrm{id})g$ using the relations from Solution 3.

References

[1] A. Connes, *Noncommutative Geometry*. Academic Press (1994).
[2] S. Majid, *Foundations of Quantum Group Theory*. Cambridge University Press (1995).
[3] S. Majid, *A Quantum Groups Primer*. London Mathematical Society, Lecture Note Series, Vol. 292. Cambridge University Press (2002).
[4] S. Woronowicz, Differential calculus on compact matrix pseudogroups (quantum groups). *Commun. Math. Phys.* **122**, 125–170 (1989).
[5] S. Majid, Noncommutative riemannian and spin geometry of the standard q-sphere. *Commun. Math. Phys.* **256**, 255–285 (2005).
[6] E. Beggs and S. Majid, Gravity induced from quantum spacetime. *Class. Quantum Grav.* **31**, 035020 (39 pp.) (2014).

Chapter 6

Mathematical Problems
of General Relativity

Juan A. Valiente Kroon

School of Mathematical Sciences
Queen Mary University of London
Mile End Road, London E1 4NS, UK
j.a.valiente-kroon@qmul.ac.uk

These notes provide an introduction to the so-called $3+1$ formulation of General Relativity. This discussion is a first necessary step towards the formulation of an initial value problem for the Einstein field equations. The basic aspects of this initial value problem, in particular the question of how to recast the field equations as a system of quasilinear wave equations and how to construct initial data for these evolution equations, are also considered.

1. Introduction

The study of the mathematical properties of the solutions to the equations of General Relativity — the Einstein field equations — has experienced a great development in recent years. Work in this area has been based on a systematic use of the so-called initial value problem for the Einstein field equations. It requires the use of ideas and techniques from various branches of Mathematics — especially Differential Geometry and Partial Differential Equations (elliptic and hyperbolic). Current mathematical challenges in the area include the analysis of the global existence of solutions to the Einstein field equations, the uniqueness of stationary black holes, the nonlinear stability of the Kerr spacetime, and the construction of initial data sets of geometrical or physical interest.

The main objective of these notes is to present a discussion of General Relativity as an initial value problem.

2.   A Brief Survey of General Relativity

General Relativity is a relativistic theory of gravity. It describes the gravitational interaction as a manifestation of the curvature of spacetime. One of the key tenets of General Relativity is that both matter and energy produce curvature of the spacetime. The way matter and energy produce curvature in spacetime is described by means of the *Einstein field equations*. One of the main properties of the gravitational field as described by General Relativity is that it can be a source of itself — this is a manifestation of the nonlinearity of the Einstein field equations. This property gives rise to a variety of phenomena that can be analysed by means of the so-called *vacuum Einstein field equations* without having to resort to any further considerations about matter sources.

As it is the case of many other physical theories, General Relativity admits a formulation in terms of an *initial value problem (Cauchy problem)* whereby one prescribes the geometry of spacetime at some instant of time and then one purports to reconstruct it from the initial data. Part of the task in the construction of the initial problem in General Relativity is to make sense of what it means to prescribe the geometry of spacetime at instant of time. A second part of the task is to show how the spacetime is to be reconstructed from the data. The initial value problem is the core of the area of research broadly known as *Mathematical Relativity* — an area of active research with a number of interesting and challenging open problems.

Before turning the attention to the initial value problem in General Relativity, it is convenient to provide a survey of the key ideas of this theory to motivate the discussion of the Cauchy problem.

2.1.   *Differential geometry*

Differential Geometry is the natural language of General Relativity. A discussion of the basic ideas of this subject goes well beyond the scope of these notes. A review in the spirit of the present notes can be found in Chapter 2 of the monograph by Valiente Kroon [1]. The various differential geometric objects will be introduced as they are required by the discussion in the text. All throughout these notes we make use the so-called *abstract index notation* whereby the tensorial character of a geometric object is encoded in the position of lowercase Latin indices. Greek indices will be used to denote *coordinate indices* in expression where a certain coordinate system has to be adopted.

2.2. The Einstein field equations

General Relativity postulates the existence of a four-dimensional manifold $\mathcal{M}$, the *spacetime manifold*, which contains events as points. This spacetime manifold is endowed with a Lorentzian metric g_{ab} which in these lectures is assumed to have signature $+2$ — i.e. $(-+++)$. By a spacetime it will understood the pair $(\mathcal{M}, g_{ab})$ where the metric g_{ab} satisfies the Einstein field equations

$$R_{ab} - \tfrac{1}{2}Rg_{ab} + \lambda g_{ab} = T_{ab}. \tag{2.1}$$

In the previous equation λ denotes the so-called *Cosmological constant* while T_{ab} is the *energy–momentum tensor* of the matter model under consideration. The main goal of *mathematical General Relativity* is to obtain a qualitative understanding of the solutions to the Einstein field equations.

The conservation of energy–momentum is encoded in the condition

$$\nabla^a T_{ab} = 0. \tag{2.2}$$

The conservation equation (2.2) is consistent with the Einstein field equations as a consequence of the second Bianchi identity. More precisely, one has the contracted Bianchi identity

$$\nabla^a \left(R_{ab} - \tfrac{1}{2}Rg_{ab} + \lambda g_{ab}\right) = 0. \tag{2.3}$$

In these notes the focus will be on systems describing *isolated bodies* so that henceforth we assume that $\lambda = 0$. Moreover, attention is restricted to the *vacuum* case for which $T_{ab} = 0$. The vacuum equations apply in the exterior region to an astrophysical source, but they usefulness is not restricted to this. There exist "stand-alone" vacuum configurations — like for example, black holes. A direct calculation in the vacuum case with vanishing Cosmological constant allows to rewrite the Einstein field equations (2.1) as (*exercise!*)

$$R_{ab} = 0. \tag{2.4}$$

The field equations prescribe the geometry of spacetime locally. However, they do not prescribe the topology of the spacetime manifold.

2.3.   Exact solutions to the Einstein field equations

What should one understand for a solution to equation (2.4)? In first instance, a solution is given by a metric g_{ab} expressed in a specific coordinate system (x^μ) — in what follows we will write this as $g_{\mu\nu}$. Exact solutions, i.e. solutions to equation (2.4) expressible in terms of elementary functions, are our main way of acquiring intuition about the behaviour of generic solutions to the Einstein field equations.

2.3.1.   The Minkowski spacetime

The simplest example of a solution to equation (2.4) is given by the metric encoded in the *Minkoswki* line element

$$g = \eta_{\mu\nu} dx^\mu dx^\nu, \quad \eta_{\mu\nu} = \mathrm{diag}(-1, 1, 1, 1). \tag{2.5}$$

One clearly verifies that for this metric $R_{\mu\nu\lambda\rho} = 0$ so that $R_{\mu\nu} = 0$ (**exercise!**). As $R_{\mu\nu}$ are the components of a tensor in a specific coordinate system one concludes then $R_{ab} = 0$. Thus, any metric related to (2.5) by a coordinate transformation is a solution to the vacuum field equation (2.4). Accordingly, one has obtained a tensor field g_{ab} satisfying equation (2.4).

The previous example shows that as a consequence of the tensorial character of the Einstein field equations a solution to the equations is, in fact, an equivalence class of solutions related to each other by means of coordinate transformations.

2.3.2.   Symmetry assumptions

In order to find further explicit solutions to equation (2.4) one needs to make some sort of assumptions about the spacetime. A standard assumption is that the spacetime has *continuous symmetries*. The notion of a continuous symmetry is formalised by the concept of a *diffeomorphism*. A diffeomorphism is a smooth map ϕ of $\mathcal{M}$ onto itself. One can think of the diffeomorphism in terms of displacements of points in the manifold along curves in the manifold — these curves are called the *orbits of the symmetry*. In what follows, let ξ^a denote the tangent vector to the orbits. The mapping ϕ is called an isometry if $\mathcal{L}_\xi g_{ab} = 0$ — that is, if the symmetry leaves the metric invariant. One can verify that this condition implies the equation (**exercise!**)

$$\nabla_a \xi_b + \nabla_b \xi_a = 0. \tag{2.6}$$

This equation is called the *Killing equation*. An important observation about this equation is that it is *overdetermined* — this means that it does not admit a solution for a general spacetime. In other words, the existence of a solution imposes restrictions on the manifold. This can be best understood by considering integrability conditions for equation (2.6). Given the commutator

$$\nabla_a \nabla_b \xi_c - \nabla_b \nabla_a \xi_c = -R^d{}_{cab} \xi_d,$$

using equation (2.6) one obtains that

$$\nabla_a \nabla_b \xi_c + \nabla_b \nabla_c \xi_a = -R^d{}_{cab} \xi_d.$$

Shuffling the indices in a cyclic way one obtains the further equations

$$\nabla_c \nabla_a \xi_b + \nabla_a \nabla_b \xi_c = -R^d{}_{bca} \xi_d, \qquad \nabla_b \nabla_c \xi_a + \nabla_c \nabla_a \xi_b = -R^d{}_{abc} \xi_d.$$

Adding the first two equations and subtracting the third one, one gets

$$2 \nabla_a \nabla_b \xi_c = (R^d{}_{abc} - R^d{}_{cab} - R^d{}_{bca}) \xi_d.$$

Finally, using the first Bianchi identity one has that $-R^d{}_{cab} = R^d{}_{abc} + R^d{}_{bca}$ so that

$$\nabla_a \nabla_b \xi_c = R^d{}_{abc} \xi_d. \tag{2.7}$$

This is an integrability condition for a solution to the Killing equation — i.e. a necessary condition that needs to be satisfied by any solution to (2.6). It shows that if one has a solution to the Killing equation, then the curvature of the spacetime is restricted.

An important type of symmetry is the so-called *spherical symmetry*. In broad terms, this means that there exists a three-dimensional group of symmetries with two-dimensional space-like orbits. Each orbit is an *homogeneous* and *isotropic* manifold. The orbits are required to be compact and to have constant positive curvature.

2.3.3. *The Schwarzschild solution*

Arguably, the most important solution to the vacuum Einstein field equations is the *Schwarzschild spacetime*, given in standard coordinates (t, r, θ, φ) by the line element

$$g = -\left(1 - \frac{2m}{r}\right) dt^2 + \left(1 - \frac{2m}{r}\right)^{-1} dr^2 + r^2(d\theta^2 + \sin^2 \theta d\varphi^2). \tag{2.8}$$

This solution is spherically symmetric and static — i.e. time independent, as it can be seen by direct inspection of the metric. In a later section we will further elaborate on the notion of static solutions. For a discussion of the interpretation and basic properties of this solution, the reader is referred to Wald's book [2]. Here we make some remarks which will motivate subsequent topics of these notes. The first one is that staticity can be obtained as a consequence of the assumption of spherical symmetry — this is usually called the *Birkhoff theorem*: any spherically symmetric solution to the vacuum field equations is locally isometric to the Schwarzschild solution (2.8). The second observation is that the Schwarzschild solution can be characterised as the only static *black hole* solution of the vacuum equations (2.4) satisfying a certain (reasonable) behaviour at infinity — *asymptotic flatness*: the requirement that asymptotically, the metric behaves like the Minkowski metric. This result is known as the *uniqueness of static black holes*. The Birkhoff and uniqueness theorems constitute examples of a type of results for solutions to the Einstein field equations known as *rigidity theorems* — these show that assuming certain properties about solutions to the Einstein field equations immediately imply other properties.

2.3.4. *Other solutions to the Einstein field equations*

A natural question is: are there other further exact solutions? The simple direct answer is in the affirmative. To obtain more solutions the natural strategy is to reduce the number of symmetries — accordingly the task of finding solutions becomes harder. A natural assumption is to look for axially symmetric and stationary solutions — stationarity is a form of time independence which is compatible with the notion of rotation. This assumption leads to the *Kerr spacetime* describing a time-independent rotating black hole. The notion of stationarity will be elaborated in a later section.

At this point, the construction of solutions by means of symmetries reaches an impasse. Although there are a huge number of explicit solutions to the Einstein field equation — see e.g. the monograph by Stephani *et al.* [3] — the number of solutions with a physical/geometric relevance is much more restricted — for a discussion of some of the physically/geometrically important solutions see e.g. the monograph by Griffiths and Podolski [4].

2.3.5. *Abstract analysis of the Einstein field equations*

An alternative to the analysis of solutions to Einstein field equations by means of the construction of exact solutions is to use the general features and structure of the equations to assert existence in an "abstract sense". This approach can be further employed to establish uniqueness and other properties of the solutions. This approach to Relativity has been strongly advocated, for example, in the monograph by Rendall [5]. After this type of analysis has been carried out, one can proceed to construct solutions numerically.

3. A First Look at the Cauchy Problem in General Relativity

A strategy to pursue the programme described in the previous paragraph is to formulate an *initial value problem* (*Cauchy problem*) for the Einstein field equations. To see what sort of issues are involved in this, it is convenient to look at a similar discussion in simpler equations.

3.1. *The initial value problem for the scalar wave equation*

In first instance, consider the wave equation $\Box \phi \equiv \nabla_a \nabla^a \phi = 0$ with respect to the metric g_{ab} of a fixed spacetime $(\mathcal{M}, g_{ab})$. In local coordinates it can be shown that (*exercise!*)

$$\Box \phi = \frac{1}{\sqrt{-\det g}} \partial_\mu (\sqrt{-\det g}\, g^{\mu\nu} \partial_\nu \phi). \tag{3.1}$$

The *principal part* of the equation $\Box \phi = 0$ corresponds to the terms in (3.1) containing the highest derivatives of the scalar field ϕ — namely $g^{\mu\nu} \partial_\mu \partial_\nu \phi$. The structure in this expression is particular of a class of partial differential equations known as *hyperbolic equations*. The prototypical hyperbolic equation is the wave equation on the Minkowski spacetime. In standard Cartesian coordinates one has that

$$\Box \phi = \eta^{\mu\nu} \partial_\mu \partial_\nu \phi = \partial_x^2 \phi + \partial_y^2 \phi + \partial_z^2 \phi - \partial_t^2 \phi = 0.$$

The Cauchy problem for the wave equations and more general hyperbolic equations is well understood in a *local setting*. Roughly speaking this means that if one prescribes the field ϕ and its derivative $\partial_\mu \phi$ at some fiduciary instant of time $t = 0$, then the equation $\Box \phi = 0$ has a solution for suitably small times (*local existence*). Moreover, this solution is *unique* in its existence interval and it has *continuous dependence* on the initial data.

The question of *global existence* is much more challenging and, in fact, an open issue for general spacetimes $(\mathcal{M}, g_{ab})$.

3.2. The Maxwell equations as wave equations

A useful model to discuss certain issues arising in the Einstein field equations are the *source-free Maxwell equations* on a fixed spacetime $(\mathcal{M}, g_{ab})$:

$$\nabla^a F_{ab} = 0, \quad \nabla_{[a} F_{bc]} = 0, \tag{3.2}$$

where $F_{ab} = -F_{ab}$ is the *Faraday tensor*. A solution to the second Maxwell equation is given by

$$F_{ab} = \nabla_a A_b - \nabla_b A_a, \tag{3.3}$$

where A_a is the so-called *gauge potential*. This statement can be verified by means of a direct computation (**exercise!**). The gauge potential does not determine the Faraday tensor in a unique way as $A_a + \nabla_a \phi$ with ϕ as scalar field gives the same F_{ab}. Substituting equation (3.3) into the first Maxwell equation one has that

$$0 = \nabla^a \left(\nabla_a A_b - \nabla_b A_a \right)$$
$$= \nabla^a \nabla_a A_b - \nabla^a \nabla_b A_a.$$

Now, using the commutator $\nabla_a \nabla_b A_c - \nabla_b \nabla_a A_c = -R^d{}_{cab} A_d$ it follows that $\nabla^a \nabla_b A_a = \nabla_b \nabla^a A_a + R^d{}_b A_d$ so that one concludes that A_b satisfies the equation (**exercise!**)

$$\nabla^a \nabla_a A_b - \nabla_b \nabla^a A_a - R^a{}_b A_a = 0. \tag{3.4}$$

The question is now: under what circumstances one can assert the existence of solutions to equation (3.4) on a smooth spacetime $(\mathcal{M}, g_{ab})$? The principal part of equation (3.4) is given by $\partial^\mu \partial_\mu A_\nu - \partial_\nu \partial^\mu A_\mu$. The key observation is that if one could remove the second term in the principal part, one would have the same principal part as for a wave equation for the components of A_a. The gauge freedom of the Maxwell equations can be exploited to this end. Making the replacement $A_\nu \to A_\nu + \nabla_\nu \phi$, with ϕ chosen such that

$$\nabla^\mu \nabla_\mu \phi = -\nabla^\mu A_\mu \tag{3.5}$$

one obtains that

$$\nabla^\mu A_\mu \to \nabla^\mu A_\mu + \nabla^\mu \nabla_\mu \phi = 0.$$

Equation (3.5) is to be interpreted as a wave equation for ϕ with source term given by $-\nabla^\mu A_\mu$. One says that the gauge potential is in the *Lorenz gauge* and it then satisfies the wave equation

$$\nabla^\mu \nabla_\mu A_\nu = R^\mu{}_\nu A_\mu. \tag{3.6}$$

In order to assert existence to the Maxwell equations one then considers the system of wave equations (3.5)–(3.6). These equations are manifestly hyperbolic so that local existence is obtained provided that suitable initial data is provided. This initial data consists of the values of ϕ, $\nabla_\mu \phi$, A_ν and $\nabla_\mu A_\nu$ at some initial time.

3.3. The Einstein field equations in wave coordinates

To provide a first discussion of the Cauchy problem for the Einstein field equations, we start by observing that given general coordinates (x^μ), the Ricci tensor R_{ab} can be explicitly written in terms of the components of the metric tensor $g_{\mu\nu}$ and its first and second partial derivatives as

$$R_{\mu\nu} = \frac{1}{2} \sum_{\lambda,\rho=0}^{3} (\partial_\lambda(g^{\lambda\rho}(\partial_\mu g_{\rho\nu} + \partial_\nu g_{\mu\rho} - \partial_\rho g_{\mu\nu})) - \partial_\nu(g^{\lambda\rho}\partial_\mu g_{\lambda\rho}))$$

$$+ \frac{1}{4} \sum_{\lambda,\rho,\sigma,\tau=0}^{3} (g^{\sigma\tau}g^{\lambda\rho}(\partial_\sigma g_{\rho\tau} + \partial_\rho g_{\sigma\tau} - \partial_\tau g_{\sigma\rho})$$

$$\times (\partial_\nu g_{\mu\lambda} + \partial_\mu g_{\lambda\nu} - \partial_\lambda g_{\mu\nu}) - g^{\rho\sigma}g^{\lambda\tau}(\partial_\nu g_{\lambda\sigma} + \partial_\lambda g_{\nu\sigma} - \partial_\sigma g_{\nu\lambda})$$

$$\times (\partial_\sigma g_{\mu\tau} + \partial_\mu g_{\sigma\tau} - \partial_\tau g_{\sigma\mu})), \tag{3.7}$$

where $g^{\lambda\rho} = (\det g)^{-1} p^{\lambda\rho}$ with $p^{\lambda\rho}$ polynomials of degree 3 in $g_{\mu\nu}$. The summation symbols have been included in the above expression for the sake of clarity. Thus, the vacuum Einstein field equation implies *second-order quasilinear partial differential equations* for the components of the metric tensor. As (3.7) is a second-order differential equation for $g_{\mu\nu}$ one may hope it is possible to recast it in the form of some type of wave equation. As it will be seen, this involves a specification of coordinates.

By recalling the formula for the Christoffel symbols in terms of partial derivatives of the metric tensor

$$\Gamma^\nu{}_{\mu\lambda} = \tfrac{1}{2}g^{\nu\rho}(\partial_\mu g_{\rho\lambda} + \partial_\lambda g_{\mu\rho} - \partial_\rho g_{\mu\lambda}),$$

and by defining $\Gamma^\nu \equiv g^{\mu\lambda}\Gamma^\nu{}_{\mu\lambda}$, one can rewrite the formula (3.7) more concisely as

$$R_{\mu\nu} = -\tfrac{1}{2}g^{\lambda\rho}\partial_\lambda\partial_\rho g_{\mu\nu} + \nabla_{(\mu}\Gamma_{\nu)} + g_{\lambda\rho}g^{\sigma\tau}\Gamma^\lambda{}_{\sigma\mu}\Gamma^\rho{}_{\tau\nu}$$
$$+ 2\Gamma^\sigma{}_{\lambda\rho}g^{\lambda\tau}g_{\sigma(\mu}\Gamma^\rho{}_{\nu)\tau}. \tag{3.8}$$

In this form, the principal part of the vacuum Einstein field equation (2.4) can be readily be identified to be given by the terms

$$-\tfrac{1}{2}g^{\lambda\rho}\partial_\lambda\partial_\rho g_{\mu\nu} + \nabla_{(\mu}\Gamma_{\nu)}.$$

An approach to the construction of systems of coordinates (x^μ) which, in turn, leads to a suitable hyperbolic equation for the components of the metric tensor g_{ab} is to require the coordinates to satisfy the equation

$$\nabla^\nu\nabla_\nu x^\mu = 0, \tag{3.9}$$

where each of the coordinates (x^μ) are treated as a scalar field over $\mathcal{M}$. A direct computation then shows that

$$\nabla_\nu x^\mu = \partial_\nu x^\mu = \delta_\nu{}^\mu,$$
$$\nabla_\lambda\nabla_\nu x^\mu = \partial_\lambda\delta_\nu{}^\mu - \Gamma^\rho{}_{\lambda\nu}\delta_\rho{}^\mu = -\Gamma^\mu{}_{\nu\lambda},$$

so that

$$\nabla^\nu\nabla_\nu x^\mu = g^{\nu\lambda}\Gamma^\mu{}_{\nu\lambda} = -\Gamma^\mu. \tag{3.10}$$

If suitable initial data is provided for equation (3.10) — the coordinate differentials dx^μ have to be chosen initially to be point-wise independent — then general theory of hyperbolic differential equations ensures the existence of a solution to equation (3.9), and as a result of equation (3.10) then $\Gamma^\mu = 0$. Thus, by a suitable choice of coordinates, the contracted Christoffel symbols Γ^μ can be locally made to vanish. This construction determines the coordinates uniquely.

Using the wave coordinates described in the previous section, equation (3.8) takes the form

$$g^{\lambda\rho}\partial_\lambda\partial_\rho g_{\mu\nu} - 2g_{\lambda\rho}g^{\sigma\tau}\Gamma^\lambda{}_{\sigma\mu}\Gamma^\rho{}_{\tau\nu} - 4\Gamma^\sigma{}_{\lambda\rho}g^{\lambda\tau}g_{\sigma(\mu}\Gamma^\rho{}_{\nu)\tau} = 0 \tag{3.11}$$

that is, one obtains a system of quasilinear wave equations for the components $(g_{\mu\nu})$ of the metric tensor g_{ab} in the coordinates (x^μ). For this system, the local Cauchy problem with appropriate data is well-posed — one can show the existence and uniqueness of solutions and their stable dependence on the data — see e.g. the survey article by Friedrich and Rendall [6]. This system of equations is called the reduced Einstein field equations. Similarly, the procedure leading to it is called a *hyperbolic reduction* of the Einstein vacuum equations. It is worth stressing that the relevance of obtaining a reduced version of the Einstein field equations in a manifestly hyperbolic form is that for these equations one readily has a developed theory of existence and uniqueness available. The introduction of a specific system of coordinates via the use of wave coordinates breaks the tensorial character of the Einstein field equations (2.4). Given a solution to the reduced Einstein field equations, the latter will also imply a solution to the actual equations as long as the coordinates (x^μ) satisfy equation (3.10) for the chosen coordinate source function appearing in the reduced equation. Thus, the standard procedure to prove local existence of solutions to the Einstein field equation with prescribed initial data is to show first the existence for a particular reduction of the equations and then prove, afterwards, that the coordinates that have been used satisfy the coordinate condition (3.10). This argument will be detailed in a subsequent section once other issues have been addressed.

The domain on which the coordinates (x^μ) form a good coordinate system depends on the initial data prescribed and the solution $g_{\mu\nu}$ itself. Since the information on $g_{\mu\nu}$ is only obtained by solving equation (3.10), there is little that can be said *a priori* about the domain of existence of the coordinates.

Finally, it is observed that the data for the reduced equation (3.11) consists of a prescription of $g_{\mu\nu}$ and $\partial_\lambda g_{\mu\nu}$ at some initial time $t = 0$.

3.4. The propagation of the wave coordinates condition

To conclude the discussion it is now shown that under suitable conditions the reduced Einstein equations imply a solution of the actual Einstein field equations. This in fact, is equivalent to showing that if the contracted Christoffel symbols $\Gamma^\mu \equiv g^{\nu\lambda}\Gamma^\mu{}_{\nu\lambda}$ vanish initially, then they also vanish at any later time.

The starting point of this discussion is the observation that the *reduced Einstein field equations* can be written as

$$R_{\mu\nu} = \nabla_{(\mu}\Gamma_{\nu)}.$$

Now, using the contracted Bianchi identity, equation (2.3), it follows that (*exercise!*)

$$\Box\Gamma_\mu + R^\nu{}_\mu\Gamma_\mu = 0.$$

This is a wave equation for the contracted Christoffel symbol. In view of its homogeneity, if

$$\Gamma_\mu = 0, \quad \nabla_\nu\Gamma_\mu = 0, \quad \text{at } t = 0, \tag{3.12}$$

then $\Gamma_\mu = 0$ at later times and accordingly $R_{\mu\nu} = 0$. That is, one has a solution to the Einstein field equations.

4. The 3 + 1 Decomposition in General Relativity

In order to understand the structure of the initial value problem in General Relativity one has to break the covariance of the theory and introduce a privileged time direction which, in turn, is used to decompose the equations of the theory.

4.1. *Submanifolds of spacetime*

Intuitively, a *submanifold* of $\mathcal{M}$, is a set $\mathcal{N} \subset \mathcal{M}$ which inherits a manifold structure from $\mathcal{M}$. The precise definition of a submanifold requires the concept of *embedding* — essentially a map $\varphi : \mathcal{N} \to \mathcal{M}$ which is injective and structure preserving; in particular, the restriction $\varphi : \mathcal{N} \to \varphi(\mathcal{N})$ is a diffeomorphism. In terms of the above concepts, a submanifold $\mathcal{N}$ is the image $\varphi(\mathcal{N}) \subset \mathcal{M}$ of a k-dimensional manifold $(k < n)$. Very often it is convenient to identify $\mathcal{N}$ with $\varphi(\mathcal{N})$.

   In what follows we will mostly be concerned with three-dimensional submanifolds. It is customary to call these *hypersurfaces*. A generic hyper-surfaces will be denoted by $\mathcal{S}$.

4.2. *Foliations of spacetime*

The presentation in this section follows very closely that of Section 2.3 in the monograph on Numerical Relativity by Baumgarte and Shapiro [7]. In what follows, we assume that the spacetime $(\mathcal{M}, g_{ab})$ is *globally hyperbolic*. That is, we assume that its topology is that of $\mathbb{R} \times \mathcal{S}$, where $\mathcal{S}$ is an orientable three-dimensional manifold. A slightly different way of saying this is that the spacetime is assumed to be *foliated* by 3-manifolds (*hypersurfaces*) $\mathcal{S}_t$, $t \in \mathbb{R}$ such that

$$\mathcal{M} = \bigcup_{t \in \mathbb{R}} \mathcal{S}_t,$$

where we identify the leaves $\mathcal{S}_t$ with $\{t\} \times \mathcal{S}$. It is assumed that the hypersurfaces $\mathcal{S}_t$ do not intersect each other. It is customary to think of the hypersurface $\mathcal{S}_0$ as an initial hypersurface on which the initial information giving rise to the spacetime is to be prescribed. Globally hyperbolic spacetimes constitute the natural class of spacetimes on which to pose an initial value problem for General Relativity.

It will be convenient to assume that the hypersurfaces $\mathcal{S}_t$ arise as the level surfaces of a scalar function t which will be interpreted as a *global time function*. From t one can define the covector $\omega_a \equiv \nabla_a t$. By construction ω_a denotes the normal to the leaves $\mathcal{S}_t$ of the foliation. The covector ω_a is *closed* — that is,

$$\nabla_{[a}\omega_{b]} = \nabla_{[a}\nabla_{b]}t = 0.$$

From ω_a one defines a scalar α called the *lapse function* via

$$g^{ab}\nabla_a t \nabla_b t = \nabla^a t \nabla_a t \equiv -\frac{1}{\alpha^2}.$$

The lapse measures how much proper time elapses between neighbouring time slices along the direction given by the normal vector $\omega^a \equiv g^{ab}\omega_b$. It is assumed that $\alpha > 0$. Notice that ω^a is assumed to be time-like so that the hypersurfaces $\mathcal{S}_t$ are space-like.

We define the unit normal n_a via $n_a \equiv -\alpha\omega_a$. The minus sign in the last definition is chosen so that n^a points in the direction of increasing t. One can readily verify that $n^a n_a = -1$. One thinks of n^a as the 4-velocity of a *normal* observer whose world line is always orthogonal to the hypersurfaces $\mathcal{S}_t$.

The spacetime metric g_{ab} induces a three-dimensional Riemannian metric h_{ij} on $\mathcal{S}_t$ — the indices $_{ij}$ are being used here to indicate that the induced metric is an intrinsically three-dimensional object. The tensors g_{ab} and h_{ij} are related to each other via $h_{ab} \equiv g_{ab} + n_a n_b$. Note that although h_{ij} is a three-dimensional object, in the previous formula spacetime indices $_{ab}$ are used — i.e. we regard the 3-metric as an object living on spacetime. One can also use h_{ab} to measure distances within $\mathcal{S}_t$. In order to see that h_{ab} is purely spatial — i.e. it has no component along n^a — one contracts with the normal:

$$n^a h_{ab} = n^a g_{ab} + n_a n^a n_b = n_b - n_b = 0.$$

The inverse 3-metric h^{ab} is obtained by raising indices with $h^{ab} = g^{ab} + n^a n^b$. The 3-metric h_{ab} can be used as a *projector tensor* which decomposes tensors into a *purely spatial part* which lies on the hypersurfaces $\mathcal{S}_t$ and a *time-like part* normal to the hypersurface. In actual computations it is convenient to consider $h_a{}^b = \delta_a{}^b + n_a n^b$. Given a tensor T_{ab} its spatial part, to be denoted by $T_{ab}^\perp$, is defined to be $T_{ab}^\perp \equiv h_a{}^c h_b{}^d T_{cd}$. One can also define a *normal projector* $N_a{}^b$ as $N_a{}^b \equiv -n_a n^b = \delta_a{}^b - h_a{}^b$. In terms of these operators an arbitrary projector can be decomposed as

$$v^a = \delta^a{}_b v^b = (h_b{}^a + N_b{}^a) v^b = v^{\perp a} - n^a n_b v^b.$$

The 3-metric h_{ij} defines in a unique manner a covariant derivative D_i — the Levi-Civita connection of h_{ij}. As in the previous paragraphs it is convenient to make use of a four-dimensional (spacetime) perspective so that we write D_a. Following this point of view one requires D_a to be torsion free and compatible with the metric h_{ab}. Taking this into account one defines for a scalar ϕ, $D_a \phi \equiv h_a{}^b \nabla_b \phi$, and, say, for a $(1,1)$-tensor $D_a T^b{}_c \equiv h_a{}^d h_e{}^b h_c{}^f \nabla_d T^e{}_f$, with an obvious extension to other tensors. Being a covariant derivative, one can naturally associate a curvature tensor $r^a{}_{bcd}$ to D_a by considering its commutator:

$$D_a D_b v^c - D_b D_a v^c = r^c{}_{dab} v^d.$$

One can verify that $r^c{}_{dab} n^d = 0$. Similarly, one can define the Ricci tensors and scalar as $r_{db} \equiv r^c{}_{dcb}$ and $r \equiv g^{ab} r_{ab}$.

4.3.   *Extrinsic curvature*

The Einstein field equation $R_{ab} = 0$ imposes some conditions on the four-dimensional Riemann tensor $R^a{}_{bcd}$. Thus, in order to understand the implications of the Einstein equations one needs to decompose $R^a{}_{bcd}$ into spatial parts. This decomposition naturally involves $r^a{}_{bcd}$, but there is more to it as this last object is purely spatial and is computed directly from h_{ab}. Hence, $r^a{}_{bcd}$ measures the *intrinsic curvature* of the hypersurface $\mathcal{S}_t$. This tensor provides no information about how $\mathcal{S}_t$ fits in $(\mathcal{M}, g_{ab})$. The missing piece of information is contained in the so-called *extrinsic curvature*.

The extrinsic curvature is defined as the following projection of the spacetime covariant derivative of the normal to $\mathcal{S}_t$:

$$K_{ab} \equiv -h_a{}^c h_b{}^d \nabla_{(c} n_{d)} = -h_a{}^c h_b{}^d \nabla_c n_d.$$

The second equality follows from the fact that n_a is *rotation free*. By construction, the extrinsic curvature is symmetric and purely spatial. It measures how the normal to the hypersurface changes from point to point. As a consequence, the extrinsic curvature also measures the rate at which the hypersurface deforms as it is carried along the normal.

A concept related to extrinsic curvature is that of the *acceleration* of a foliation $a_a \equiv n^b \nabla_b n_a$. Using $n^d \nabla_c n_d = 0$, one can compute

$$K_{ab} = -h_a{}^c h_b{}^d \nabla_c n_d = -(\delta_a{}^c + n_a n^c)(\delta_b{}^d + n_b n^d)$$
$$= -(\delta_a{}^c + n_a n^c)\delta_b{}^d \nabla_c n_d = -\nabla_a n_b - n_a a_b.$$

An alternative expression of the extrinsic curvature is given in terms of the Lie derivative. To obtain this one computes

$$\mathcal{L}_n h_{ab} = \mathcal{L}_n(g_{ab} + n_a n_b) = 2\nabla_{(a} n_{b)} + n_a \mathcal{L}_n n_b + n_b \mathcal{L}_n n_a$$
$$= 2(\nabla_{(a} n_{b)} + n_{(a} a_{b)}) = -2K_{ab}.$$

The so-called *mean curvature* is defined as the trace $K \equiv g^{ab} K_{ab} = h^{ab} K_{ab}$.

4.4. The Gauss–Codazzi and Codazzi–Mainardi equations

Given the extrinsic curvature of an hypersurface $\mathcal{S}$, we now look how this relates to the curvature of spacetime. A computation using the definitions of the previous section shows that

$$D_a D_b v^c = h_a{}^p h_b{}^q h_r{}^c \nabla_p \nabla_q v^r - K_{ab} h_r{}^c n^p \nabla_p v^r - K_a{}^c K_{bp} v^p.$$

Combining with the commutator $D_a D_b v^c - D_b D_a v^c = r^c{}_{dab} v^d$, after some manipulations one obtains (***exercise!***)

$$r_{abcd} + K_{ac} K_{bd} - K_{ad} K_{cb} = h_a{}^p h_b{}^q h_c{}^r h_d{}^s R_{pqrs}. \tag{4.1}$$

This equation is called the *Gauss–Codazzi equation*. It relates the spatial projection of the spacetime curvature tensor to the three-dimensional curvature.

A further important identity arises from considering projections of R_{abcd} along the normal direction. This involves a spatial derivative of the extrinsic curvature. Namely, $D_a K_{bc} = h_a{}^p h_b{}^q h_c{}^r \nabla_p K_{qr}$. From this expression after some manipulations one can deduce (***exercise!***)

$$D_b K_{ac} - D_a K_{bc} = h_a{}^p h_b{}^q h_c{}^r n^s R_{pqrs}. \qquad (4.2)$$

This equation is called the *Codazzi–Mainardi equation*.

In the sequel, we explore the consequences of equations (4.1) and (4.2) for the initial value problem in General Relativity.

4.5.   *The constraint equations of General Relativity*

The $3 + 1$ decomposition of the Einstein field equations allows to identify the intrinsic metric and the extrinsic curvature of an initial hypersurface $\mathcal{S}_0$ as the initial data to be prescribed for the evolution equations of General Relativity.

In what follows, we will make use of the Gauss–Codazzi and the Codazzi–Mainardi equations to extract the consequences of the vacuum Einstein field equations $R_{ab} = 0$ on a hypersurface $\mathcal{S}$. Contracting the Gauss–Codazzi equation (4.1) one finds that

$$h^{pr} h_b{}^q h_d{}^s R_{pqrs} = r_{bd} + K K_{bd} - K^c{}_d K_{cb},$$

where $K \equiv h^{ab} K_{ab}$ denotes the trace of the extrinsic curvature. A further contraction yields $h^{pr} h^{qs} R_{pqrs} = r + K^2 - K_{ab} K^{ab}$. Now, the left-hand side can be expanded into

$$\begin{aligned} h^{pr} h^{qs} R_{pqrs} &= (g^{pr} + n^p n^s)(g^{qs} + n^q n^s) R_{pqrs} \\ &= R + 2 n^p n^r R_{pr} + n^p n^r n^q n^s R_{pqrs} = 0. \end{aligned}$$

The last term vanishes because of the symmetries of the Riemann tensor. Combining the equations from the previous calculations one obtains the so-called *Hamiltonian constraint* (***exercise!***):

$$r + K^2 - K_{ab} K^{ab} = 0. \qquad (4.3)$$

One can proceed in a similar way with the Codazzi–Mainardi equation (4.2). Contracting once one has that $D^b K_{ab} - D_a K = h_a{}^p h^{qr} n^s R_{pqrs}$. The right-hand side of this equation can be, in turn, expanded as

$$\begin{aligned} h_a{}^p h^{qr} n^s R_{pqrs} &= -h_a{}^p (g^{qr} + n^p n^r) n^s R_{qprs} \\ &= -h_a{}^p n^s R_{ps} - h_a{}^p n^q n^r n^s R_{pqrs} = 0, \end{aligned}$$

where in the last equality one makes use, again, of the vacuum equations and the symmetries of the Riemann tensor. Combining the previous expressions one obtains the so-called *momentum constraint* (***exercise!***)

$$D^b K_{ab} - D_a K = 0. \tag{4.4}$$

The Hamiltonian and momentum constraint equations (4.3) and (4.4) involve only the three-dimensional intrinsic metric, the extrinsic curvature and their spatial derivatives. They are the conditions that allow a three-dimensional slice with data (h_{ab}, K_{ab}) to be embedded in a four-dimensional spacetime $(\mathcal{M}, g_{ab})$. The existence of the constraint equations implies that the data for the Einstein field equations cannot be prescribed freely. The nature of the constraint equations and possible procedures to solve them will be analysed later in these notes.

An important point still to be clarified is whether the fields h_{ij} and K_{ij} indeed correspond to data for the Einstein field equations. To see this, one has to analyse the evolution equations implied by the Einstein field equations.

4.6. *The evolution equations of General Relativity*

To discuss the evolution equations of General Relativity one needs a further geometric identity — the so-called *Ricci equation*. To obtain this, one computes

$$\mathcal{L}_n K_{ab} = n^c \nabla_c K_{ab} + 2K_{c(a} \nabla_{b)} n^c$$
$$= -n^c \nabla_c \nabla_a n_b - n^c \nabla_c (n_a a_b) - 2K_{c(a} K_{b)}{}^c - 2K_{c(a} n_{b)} a^c.$$

Now, making use of the commutator $\nabla_c \nabla_a n_b - \nabla_a \nabla_c n_b = R_{dbac} n^d$, one obtains

$$\mathcal{L}_n K_{ab} = -n^d n^c R_{dbac} - n^c \nabla_a \nabla_c n_b - n^c a_b \nabla_c n_a - n^c n_a \nabla_c a_b$$
$$- 2K^c{}_{(a} K_{b)c} - 2K_{c(a} n_{b)} a^c.$$

Furthermore, using $a_b = n^c \nabla_c n_b$ and

$$n^c \nabla_a \nabla_c n_b = \nabla_a a_b - (\nabla_a n^c)(\nabla_c n_b) = \nabla_a a_b - K_a{}^c K_{cb} - n_a a^c K_{cb},$$

after some cancellations one gets

$$\mathcal{L}_n K_{ab} = -n^d n^c R_{dbac} - \nabla_a a_b - n^c n_a \nabla_c a_b - a_a a_b$$
$$- K^c{}_b K_{ac} - K_{ca} n_b a^c. \tag{4.5}$$

It is observed that $\mathcal{L}_n K_{ab}$ is a spatial object in the sense that $n^a \mathcal{L}_n K_{ab} = 0$ (***exercise!***). This means that in equation (4.5) one can project the free indices to obtain

$$\mathcal{L}_n K_{ab} = -n^d n^c h_a{}^q h_b{}^r R_{drqc} - h_a{}^q h_b{}^r \nabla_q a_r - a_a a_b - K_b{}^c K_{ac}.$$

Finally, using the identity (***exercise!***)

$$D_a a_b = -a_a a_b + \frac{1}{\alpha} D_a D_b \alpha,$$

some further simplifications yield the desired *Ricci equation* (***exercise!***):

$$\mathcal{L}_a K_{ab} = n^d n^c h_a{}^q h_b{}^r R_{drcq} - \frac{1}{\alpha} D_a D_b \alpha - K_b{}^c K_{ac}. \tag{4.6}$$

Geometrically, this equation relates the derivative of the extrinsic curvature in the normal direction to an hypersurface $\mathcal{S}$ to a time projection of the Riemann tensor.

The discussion from the previous paragraphs suggests that the Einstein field equations will imply an *evolution* of the data (h_{ab}, K_{ab}). Previously, it has been assumed that the spacetime $(\mathcal{M}, g_{ab})$ is foliated by a time function t whose level surfaces correspond to the leaves of the foliation. Recalling that $\omega_a = \nabla_a t$, we consider now a vector t^a (the *time vector*) such that

$$t^a = \alpha n^a + \beta^a, \qquad \beta_a n^a = 0. \tag{4.7}$$

The vector β^a is called the *shift vector*. The time vector t^a will be used to *propagate coordinates* from one time slice to another. In other words, t^a connects points with the same spatial coordinate — hence, the shift vector measures the amount by which the spatial coordinates are shifted within a slice with respect to the normal vector. Together, the lapse and shift determine how coordinates evolve in time. The choice of these functions is fairly arbitrary and, hence, they are known as *gauge functions*. The lapse function reflects the freedom to choose the sequence of time slices, pushing them forward by different amounts of proper time at different spatial points on a slice — this idea is usually called the *many-fingered nature of time*. The shift vector reflects the freedom to relabel spatial coordinates on each slices in an arbitrary way. Observers *at rest* relative to the slices follow the normal congruence n^a and are called *Eulerian observers*, while observers following the congruence t^a are called *coordinate observers*.

It is observed that as a consequence of expression (4.7) one has $t^a \nabla_a t = 1$ so that the integral curves of t^a are naturally parametrised by t.

Recalling that $K_{ab} = -\frac{1}{2}\mathcal{L}_n h_{ab}$ and using equation (4.7) one concludes that (*exercise!*)

$$\mathcal{L}_t h_{ab} = -2\alpha K_{ab} + \mathcal{L}_\beta h_{ab}, \tag{4.8}$$

where it has been used that $\mathcal{L}_t h_{ab} = \mathcal{L}_{\alpha n + \beta} h_{ab} = \alpha \mathcal{L}_n h_{ab} + \mathcal{L}_\beta h_{ab}$. Equation (4.8) will be interpreted as an evolution equation for the intrinsic metric h_{ab}. To construct a similar equation for the extrinsic curvature one makes use of the Ricci equation (4.6). It is observed that

$$n^d n^c h_a{}^q h_b{}^r R_{drcq} = h^{cd} h_a{}^q h_b{}^r R_{drcq} - h_a{}^q h_b{}^r R_{rq}$$

$$= h^{cd} h_a{}^q h_b{}^r R_{drcq},$$

where to obtain the second equality the vacuum Einstein field equations $R_{ab} = 0$ have been used. The remaining term, $h^{cd} h_a{}^q h_b{}^r R_{drcq}$, is dealt with the Gauss–Codazzi equation (4.1). Finally, using that

$$\mathcal{L}_t K_{ab} = \mathcal{L}_{\alpha n + \beta} K_{ab} = \alpha \mathcal{L}_n K_{ab} + \mathcal{L}_\beta K_{ab},$$

one concludes that (*exercise!*)

$$\mathcal{L}_t K_{ab} = -D_a D_b \alpha + \alpha(r_{ab} - 2K_{ac}K^c{}_b + KK_{ab}) + \mathcal{L}_\beta K_{ab}. \tag{4.9}$$

This is the desired evolution equation for K_{ab}. Equations (4.8) and (4.9) are called the *ADM evolution equations*. They determine the evolution of the data (h_{ab}, K_{ab}). Together with the constraint equations (4.3) and (4.4) they are completely equivalent to the vacuum Einstein field equations.

Remark. The evolution equations (4.8) and (4.9) are first-order equations — compare with the wave equation for the components of the metric g_{ab} discussed in Section 3.3. However, the equations are not hyperbolic! Thus, one cannot apply directly the standard PDE theory to assert existence of solutions. Nevertheless, there are some more complicated versions of them which do have the property of hyperbolicity.

4.7. The 3 + 1 form of the spacetime metric

The discussion of the evolution equations given in the previous section has been completely general. By this we mean that the only assumption that has been made about the spacetime is that it is globally hyperbolic so that a foliation and a corresponding time vector exist. The discussion of the 3 + 1 decomposition can be further particularised by introducing adapted coordinates. In this section we briefly discuss how this can be done.

Firstly, it is recalled that the hypersurfaces of the foliation of a space-time $(\mathcal{M}, g_{ab})$ can be given as the level surfaces of a time function t. Now, we already have seen that $\nabla_a t^a = 1$. The latter combined with $\nabla_a t = (1, 0, 0, 0)$ readily gives that $t^\mu = (1, 0, 0, 0)$. The latter implies that the Lie derivative along the direction of t^a is simply a partial derivative — that is, $\mathcal{L}_t = \partial_t$. Clearly, from the previous discussion it also follows that the spatial components of the unit normal must vanish — i.e. one has that $n_i = 0$. Since the contraction of spatial vectors with the normal must vanish, it follows that all components of spatial tensors with a contravariant index equal to zero must vanish. For the shift vector one has that $n_a \beta^a = n_0 \beta^0 = 0$ so that $\beta^\mu = (0, \beta^\gamma)$. Since one has that $t^a = \alpha n^a + \beta^a$, it follows then that $n^\mu = (\alpha^{-1}, -\alpha^{-1}\beta^\gamma)$. Moreover, from the normalisation condition $n_a n^a = -1$ one finds $n_\mu = (-\alpha, 0, 0, 0)$. Now, recalling that $h_{ab} = g_{ab} + n_a n_b$ one concludes that $h_{\alpha\beta} = g_{\alpha\beta}$. In these *adapted coordinates* the 3-metric of the hypersurfaces of the foliation are simply the spatial part of the space-time metric g_{ab}. Moreover, since the time components of spatial contravariant tensors have to vanish, one also has that $h^{\mu 0} = 0$. One concludes that one can write

$$g^{\mu\nu} = h^{\mu\nu} - n^\mu n^\nu = \begin{pmatrix} -\alpha^{-2} & \alpha^{-2}\beta^\gamma \\ \alpha^{-2}\beta^\delta & h^{\gamma\delta} - \alpha^{-2}\beta^\gamma\beta^\delta \end{pmatrix}.$$

This last expression can be inverted to yield

$$g_{\mu\nu} = \begin{pmatrix} -\alpha^2 + \beta_\gamma\beta^\gamma & \beta_\gamma \\ \beta_\delta & h_{\gamma\delta} \end{pmatrix},$$

where $\beta_\gamma \equiv h_{\gamma\delta}\beta^\delta$. Alternatively, one has that

$$g = -\alpha^2 dt^2 + h_{\gamma\delta}(\beta^\gamma dt + dx^\gamma)(\beta^\delta dt + dx^\delta). \qquad (4.10)$$

The latter is known as the 3 + 1 *form* of the spacetime metric.

4.7.1.   *An example: The Schwarzschild spacetime*

As we have already discussed, the metric of the Schwarzschild spacetime can be expressed in standard coordinates in terms of the line element (2.8). This form of the metric is not the best one for a 3 + 1 decomposition of the spacetime. Instead, it is better to introduce an *isotropic radial coordinate* $\bar{r}$ via

$$r = \bar{r}\left(1 + \frac{m}{2\bar{r}}\right)^2.$$

In terms of the later one obtains the line element of the Schwarzschild spacetime in *isotropic coordinates* (***exercise!***):

$$g = -\left(\frac{1 - m/2\bar{r}}{1 + m/2\bar{r}}\right)^2 dt^2 + \left(1 + \frac{m}{2\bar{r}}\right)^4 (d\bar{r}^2 + \bar{r}^2 d\theta^2 + \bar{r}^2 \sin^2\theta d\varphi).$$

The normal $\omega_a = \nabla_a t$ is then readily given by $\omega_\mu = (1, 0, 0, 0)$. Thus, one readily reads the lapse function to be (***exercise!***)

$$\alpha = \frac{1 - m/2\bar{r}}{1 + m/2\bar{r}}.$$

The spatial metric is then (***exercise!***)

$$h = \left(1 + \frac{m}{2\bar{r}}\right)^4 (d\bar{r}^2 + \bar{r}^2 d\theta^2 + \bar{r}^2 \sin^2\theta d\varphi).$$

One also notices that the shift vanishes — i.e. one has that $\beta^i = 0$. Since $\beta^\alpha = 0$ and $h_{\alpha\beta}$ is independent of time, one readily finds that the extrinsic curvature vanishes $K_{\alpha\beta} = 0$ (***exercise!***).

The isotropic form of the Schwarzschild metric yields a foliation of spacetime that follows the static symmetry of the spacetime. In this foliation, the intrinsic 3-metric of the leaves does not evolve. Any other foliation not aligned with the static Killing vector will give rise to a non-trivial evolution for both $h_{\alpha\beta}$ and $K_{\alpha\beta}$.

5. A Closer Look at the Constraint Equations

The purpose of this section is to explore some aspects of the constraint equations, and in particular, the manner one could expect to be able to solve them.

As shown in Section 4.5, the Einstein field equations imply the constraint equations (4.3) and (4.4) on a (spatial) hypersurface S. These equations constrain the possible choices of pairs (h_{ij}, K_{ij}) corresponding to initial data to the Einstein field equations. The constraint equations are intrinsic equations, that is, they only involve objects which are defined on the hypersurface S without any further reference to the "bulk" of the spacetime $(\mathcal{M}, g_{ab})$.

The Einstein constraint constitute a coupled, nonlinear system of equations for (h_{ij}, K_{ij}). The main difficulty in constructing an solution to the equations lies in the fact that the equations are an underdetermined system: one has four equations for 12 unknowns — the independent components

of two symmetric spatial tensors. Even exploiting the coordinate freedom
to "kill off" three components of the tensors, one is still left with nine
unknowns. This feature indicates that there should be some freedom in the
specification of *data* for the equations. The task is to identify what this free
data is.

To render the problem manageable, we make a standard simplifying
assumption and consider initial data sets for which $K_{ij} = 0$ everywhere
on $\mathcal{S}$. This class of initial data are called *time symmetric*. The reason for
this name is that if $K_{ij} = 0$ at $\mathcal{S}$ then the evolution equations imply that
$\partial_t h_{ij} = 0$ on $\mathcal{S}$. This equation is invariant under the replacement $t \mapsto -t$. It
follows that the resulting spacetime has a reflection symmetry with respect
to the hypersurface $\mathcal{S}$ which can be regarded as a *moment of time symmetry*.

If $K_{ij} = 0$ everywhere on $\mathcal{S}$ then the momentum constraint is automatically solved, and the Hamiltonian constraint reduces to $r = 0$. That is,
the initial 3-metric has to be such that its Ricci scalar vanishes — notice
that this does not mean that the hypersurface is flat! Still, the time symmetric Hamiltonian constraint, regarded as an equation for h_{ij}, is highly
nonlinear. Moreover, one still has six unknowns and one equation — even
choosing coordinates, one still is left with three unknowns. Now, clearly for
an arbitrary metric $\bar{h}_{ij}$ one has that $\bar{r} \neq 0$. An idea to solve the constraint is
then to introduce a factor that compensates this. This idea leads naturally
to the notion of conformal transformations. Two metrics h_{ij}, $\bar{h}_{ij}$ are said
to be conformally related if there exists a positive scalar ϑ (the *conformal
factor*) such that

$$h_{ij} = \vartheta^4 \bar{h}_{ij}. \tag{5.1}$$

The power of ϑ used in the above equation leads to simple equations in
3-dimensions. When discussing conformal transformations on dimensions
$n \geq 4$, other powers may be more useful. The metric $\bar{h}$ will be called
the *background* metric. Loosely speaking, the conformal factor absorbs the
overall scale of the metric.

Given the conformal transformation (5.1), it is important to analyse
its effects on other geometric objects. In particular, recall that the three-
dimensional Christoffel symbols are given by

$$\gamma^{\alpha}{}_{\beta\gamma} = \tfrac{1}{2} h^{\alpha\delta} (\partial_\beta h_{\gamma\delta} + \partial_\gamma h_{\beta\delta} - \partial_\delta h_{\beta\gamma}).$$

Substituting (5.1) into the previous equation one finds after some calculations that

$$\gamma^{\alpha}{}_{\beta\gamma} = \bar{\gamma}^{\alpha}{}_{\beta\gamma} + 2(\delta_\beta{}^\alpha \partial_\gamma \ln \vartheta + \delta_\gamma{}^\alpha \partial_\beta \ln \vartheta - \bar{h}_{\beta\gamma} \bar{h}^{\alpha\delta} \partial_\delta \ln \vartheta),$$

where $\bar\gamma^\alpha{}_{\beta\gamma}$ denote the Christoffel symbols for the metric coefficients $\bar h_{\alpha\beta}$ and it has been used that $h^{ij} = \vartheta^{-4}\bar h^{ij}$ (*exercise!*). A lengthier computation yields the following transformation law for the three-dimensional Ricci tensor (*exercise!*):

$$r_{ij} = \bar r_{ij} - 2(\bar D_i \bar D_j \ln\vartheta + \bar h_{ij}\bar h^{lm}\bar D_l \bar D_m \ln\vartheta)$$
$$+ 4(\bar D_i \ln\vartheta \bar D_j \ln\vartheta - \bar h_{ij}\bar h^{lm}\bar D_l \ln\vartheta \bar D_m \ln\vartheta).$$

Furthermore, one has that (*exercise!*)

$$r = \vartheta^{-4}\bar r - 8\bar\theta^{-5}\bar D_k \bar D^k \vartheta.$$

In the above expressions $\bar D$ denotes the covariant derivative of the background metric $\bar h_{ij}$.

Using $r = 0$ in the transformation law for the Ricci scalar given above, one finds that

$$\bar D_k \bar D^k \vartheta - \tfrac{1}{8}\bar r\vartheta = 0. \tag{5.2}$$

In Geometric Analysis, this equation is sometimes called the *Yamabe equation*. Given a fixed background metric $\bar h_{ij}$, equation (5.2) can be read as a differential condition for the conformal factor ϑ. Given a solution ϑ, one has by construction that $h_{ij} = \vartheta^4 \bar h_j$ such that $r = 0$ and one has constructed a solution to the time symmetric Einstein constraints. The Yamabe equation is *elliptic*: the operator $\bar D_k \bar D^k$ is the *Laplacian operator* associated to the metric $\bar h_{ij}$ — if $\bar h_{ij} = \delta_{ij}$, the flat metric in Cartesian coordinates, then

$$\bar D_k \bar D^k = \delta^{\alpha\beta}\partial_\alpha\partial_\beta = \partial_x^2 + \partial_y^2 + \partial_z^2.$$

Given a linear second-order elliptic equation like (5.2), appropriate boundary conditions ensure the existence of a unique solution on $\mathcal{S}$.

Following the discussion of the previous paragraph, choose the flat metric as background metric. That is, let $\bar h_{\alpha\beta} = \delta_{\alpha\beta}$. In this case, the metric $h_{\alpha\beta} = \vartheta^4 \delta_{\alpha\beta}$ is said to be *conformally flat*. Conformal flatness provides a simplified setting to carry out calculations. In particular, one has that $\bar r = 0$ so that the Yamabe equation reduces to the *flat Laplace equation*

$$\bar D_k \bar D^k \vartheta = 0.$$

In the discussion of isolated systems (i.e. astrophysical sources) one is interested in solutions which are *asymptotically flat*. That is, one requires $\vartheta = 1 + O(r^{-1})$, for $r \to \infty$, where $r^2 = x^2 + y^2 + z^2$ is the standard radial coordinate. Solutions to the Laplace equation with the above asymptotic behaviour are well known. In particular, a *spherically symmetric* solution is given by $\vartheta = 1 + m/2r$ where m is a constant. This solution to the time symmetric constraints is the 3-metric of the Schwarzschild spacetime in isotropic coordinates:

$$h = \left(1 + \frac{m}{2r}\right)^4 (dr^2 + r^2 d\theta^2 + r^2 \sin^2\theta d\varphi^2).$$

The above 3-metric is singular at $r = 0$ — this singularity is a coordinate artefact. By considering the coordinate inversion $r = m^2/4\bar{r}$ it can be seen that the metric transforms into

$$h = \left(1 + \frac{m}{2\bar{r}}\right)^4 (d\bar{r}^2 + \bar{r}^2 d\theta^2 + \bar{r}^2 \sin^2\theta d\varphi^2).$$

The inversion transforms the metric into itself — that is, it is a discrete isometry. In particular, one has that the point $r = 0$ is mapped to infinity. Thus, the metric is perfectly regular everywhere and $r = 0$ is, in fact, the infinity of an asymptotically flat region. The hypersurface $\mathcal{S}$ has a non-trivial topology — it corresponds to a *wormhole*, see Fig. 1, left. The radius given by $r = m/2$ corresponds to the minimum areal radius — this is called the *throat* of the black hole. Observe that $r = m/2 = \bar{m}/2$. The throat corresponds to the intersection of the black hole horizon with the hypersurface $\mathcal{S}$. The inversion reflects points with respect to the throat.

The construction described in the previous paragraphs can be extended to include an arbitrary number of black holes. This is made possible by the linearity of the flat Laplace equation. Indeed, one can consider the conformal factor

$$\vartheta = 1 + \frac{m_1}{2r_1} + \frac{m_2}{2r_2} \tag{5.3}$$

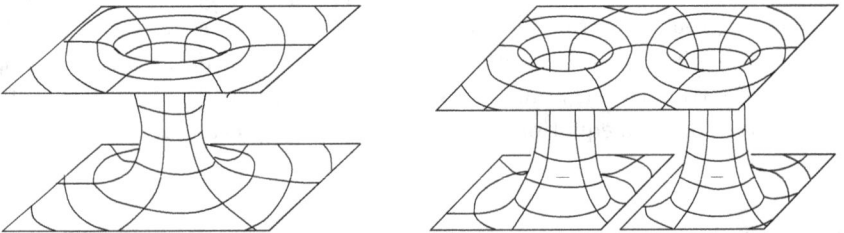

Fig. 1.   Left: embedding diagram of time-symmetric Schwarzschild data. Right: embedding diagram of time-symmetric Brill–Lindquist data.

with $r_1 = |x^i - x_1^i|$ and $r_2 = |x^i - x_2^i|$, and where x_1^i and x_2^i denote the (fixed) location of two black holes with *bare masses* m_1 and m_2. The solution to the constraint equations given by the conformal factor (5.3) is called the *Brill–Lindquist solution* [8]. It describes a pair of black holes instantaneously at rest at a moment of time symmetry. This solution is used as initial data to simulate the head-on collision of two black holes. Each throat connects to its own asymptotically flat region. The drawing of the corresponding three-dimensional manifold gives three different sheets, each corresponding to a different asymptotically flat region — see Fig. 1, right.

More complicated solutions to the constraint equations can be obtained by including a non-vanishing extrinsic curvature. In this way one can provide data for a rotating black hole or even a pair of rotating black holes spiralling around each other. The constraint equations in these cases have to be solved numerically.

6. Time-Independent Solutions

In this section we provide a brief discussion of time-independent solutions to the Einstein field equations. These solutions provide a valuable source of intuition and insight for the analysis of dynamic spacetimes.

6.1. *The time-independent wave equation*

Before analysing the Einstein field equations, it is useful to look at simpler model equations. To this end, consider a scalar field on the Minkowski spacetime satisfying the wave equation $(\Delta - \partial_t^2)\phi = 0$, where Δ denotes the flat three-dimensional Laplacian. Now, consider time-independent solutions — i.e. such that $\partial_t \phi = 0$. It follows that $\Delta \phi = 0$. The key observation is that an equation which is originally hyperbolic becomes elliptic under the assumption of time independence. This is a generic feature that can be observed in other theories — like the Maxwell equations and the Einstein field equations.

The energy of the scalar field at some time t is given by

$$E(t) = \int_{\mathcal{S}_t} \left((\partial_t \phi)^2 + |D\phi|^2 \right) d^3 x.$$

In order to have finiteness of the energy one needs the boundary conditions $\phi(t, x^i)$, $\partial_t \phi(t, x^i) \to \infty$ as $|x| \to \infty$.

An important difference between hyperbolic equations and elliptic ones is that while in the former, properties of solutions can be localised and perturbations have finite propagation speed, for the latter the properties of solutions are global. For example, if $\phi = O(1/r)$ as $r \to \infty$ and $\Delta\phi = 0$, then it follows that $\phi = 0$. This follows from the integral

$$0 = \int_{\mathbb{R}^3} \phi \Delta\phi dx^3 = \int_{\mathbb{R}^3} |D\phi|^2 dx^3,$$

where the Green's identity has been used. It follows that $|D\phi|^2 = 0$ everywhere on $\mathbb{R}^3$ so that ϕ is constant. Due to the decay conditions, ϕ must necessarily vanish. This type of argument will be used repeatedly for the Einstein equations. In order to avoid the vanishing of ϕ in this case, one needs to consider the inhomogeneous problem — that is, one needs to consider sources.

6.2. *Time independence: Stationarity and staticity*

Time independence is imposed in General Relativity by requiring on the spacetime $(\mathcal{M}, g_{ab})$ the existence of a time-like Killing vector ξ^a — the spacetime is then said to be *stationary*. If, in addition, the Killing vector is hypersurface orthogonal — i.e. it is the gradient of some scalar function — then one says that ξ^a is a *static Killing vector*. The Schwarzschild and Kerr solutions are, respectively, static and stationary. Stationary solutions to the Einstein field equations allow for the possibility of rotating gravitational fields.

Let n_a denote the unit normal of an hypersurface $\mathcal{S}$. Now, if $\xi^a n_a = 0$, i.e. the Killing vector is orthogonal to $\mathcal{S}$, then a calculation shows that (*exercise!*)

$$\xi_{[a} \nabla_b \xi_{c]} = 0. \tag{6.1}$$

The latter condition characterises *hypersurface orthogonality* — that is, a Killing vector is hypersurface orthogonal if and only if (6.1) holds. The proof of this result is classical and can be found in various textbooks — e.g. the textbook by Wald [2].

In a static spacetime, it is natural to choose adapted coordinates such that $\xi^\mu \partial_\mu = \partial_t$ — that is, the time coordinate is adapted to the flow lines of the Killing vector. Now, using the Killing vector condition $\mathcal{L}_\xi g_{ab} = 0$ and the definitions of $h_{\alpha\beta}$ and $K_{\alpha\beta}$ one can show that (*exercise!*)

$$\partial_t h_{\alpha\beta} = \partial_t K_{\alpha\beta} = 0.$$

In what follows we will analyse the simplifications introduced in the $3 + 1$ decomposed line element of equation (4.10) by the assumption of staticity and the use of an adapted time coordinate. If the Killing vector is hypersurface orthogonal then the Killing vector has to be proportional to the normal to the hypersurface $\mathcal{S}$. That is, one has $\xi_\mu = \alpha \nabla_\mu t$. However, the Killing vector can be decomposed in a lapse and a shift part as $\xi^a = \alpha n^a + \beta^a$. Comparing both expressions one finds that $\beta^\alpha = 0$. Thus, one has that

$$g = -\alpha^2 dt^2 + h_{\alpha\beta} dx^\alpha dx^\beta,$$

with $h_{\alpha\beta}$ time independent. The time evolution equation for $h_{\alpha\beta}$ then takes the form $\partial_t h_{\alpha\beta} = -2\alpha K_{\alpha\beta} = 0$. Now, as the lapse cannot vanish one has that $K_{\alpha\beta} = 0$. That is, the hypersurfaces of the foliation adapted to the static Killing vector have no extrinsic curvature — this property is preserved as, we already have seen that $\partial_t K_{\alpha\beta} = 0$.

It follows from the discussion in the previous paragraphs that vacuum static solutions to the Einstein equations are characterised solely in terms of the lapse α and the 3-metric h_{ij}. In order to obtain equations for these quantities one considers the Hamiltonian constraint, equation (4.3), and the evolution equation for K_{ij}, equation (4.9). Setting $K_{\alpha\beta} = \partial_t K_{\alpha\beta} = 0$ readily yields

$$D_i D_j \alpha = r_{ij}, \quad r = 0, \tag{6.2}$$

where, as before, r denotes the Ricci tensor of the 3-metric h_{ij}. These equations are known as the *static vacuum Einstein equations*.

6.3. *Exploring the static equations*

As a first example of the content and implications of the static equations (6.2), let $\mathcal{S} \approx \mathbb{R}^3$ — i.e. the hypersurface $\mathcal{S}$ has the topology of Euclidean space. Suppose that the fields α and h_{ij} decay at infinity in such a way that

$$\alpha \to 1, \quad h_{\alpha\beta} - \delta_{\alpha\beta} \to 0, \quad \text{as } |x| \to \infty.$$

The first condition implies that the Killing vector behaves asymptotically like the static Killing vector of Minkowski spacetime. The second condition means that the 3-metic is assumed to be asymptotically flat (Euclidean) at infinity. Taking the trace of the first equation in (6.2) and then using the second equation it follows that $\Delta\alpha = D_k D^k \alpha = 0$. Now, consider

$$0 = \int_{\mathcal{S}} \alpha \Delta \alpha d^3 x = \int_{\mathcal{S}} |D\alpha|^2 d^3 x,$$

again, as a consequence of the Gauss theorem. Thus, $|D\alpha|^2 = h^{ij}D_i\alpha D_j\alpha = 0$. Hence, α is a constant. Using the asymptotic condition $\alpha \to 1$ it follows $\alpha = 1$ everywhere. Using the first equation in (6.2) one concludes that $r_{ij} = 0$. Now, in 3-dimensions the Ricci tensor determines fully the curvature of the manifold. Accordingly, one concludes $r_{ijkl} = 0$, so that $h_{\alpha\beta} = \delta_{\alpha\beta}$ — the Euclidean flat metric. The line element we have obtained is then

$$g = -dt^2 + \delta_{\alpha\beta}dx^\alpha dx^\beta.$$

This solution is the Minkowski spacetime! This result is known as *Licnerowicz's theorem*.

Theorem 6.1. *The only globally regular static solution to the Einstein equations with S having trivial topology (i.e. $S \approx \mathbb{R}^3$) and such that*

$$\alpha \to 1, \quad h_{\alpha\beta} - \delta_{\alpha\beta} \to 0, \quad as \ |x| \to \infty$$

is the Minkowski spacetime.

The above theorem demonstrates the rigidity of the Einstein field equations. In order to obtain more interesting regular solutions, one requires either some matter sources or a non-trivial topology for S as in the case of the Schwarzschild spacetime — recall the Einstein–Rosen bridge! The result can be interpreted as a first, very basic uniqueness black hole result. If one wants to have a black hole solution one needs non-trivial topology!

7. Symmetries and the Initial Value Problem

An issue which often arises in the analysis of the Cauchy problem for the Einstein field equations is that of encoding the existence of a spacetime symmetry in the initial data for the Einstein field equations. This naturally leads to the notion of *Killing initial data (KID)*.

To analyse the question raised in the previous paragraph, it is necessary to first consider some consequences of the Killing equation $\nabla_a\xi_b + \nabla_b\xi_a = 0$. Applying ∇^a to this equation and commuting covariant derivatives one finds that

$$0 = \nabla^a\nabla_a\xi_b + \nabla^a\nabla_b\xi_a = \Box\xi_b + \nabla_b\nabla^a\xi_a + R^c{}_a{}^a{}_b\xi_c$$

$$= \Box\xi_b - R^c{}_b\xi_c,$$

where it has been used that $\nabla^a \xi_a = 0$. Accordingly, in vacuum one has that a Killing vector satisfies the wave equation

$$\Box \xi_a = 0. \tag{7.1}$$

This equation is an integrability condition for the Killing equation. Notice however, that not every vector solution to the wave equation (7.1) is a Killing vector. A vector ξ_a satisfying equation (7.1) will be called a *Killing vector candidate*.

Now, in what follows let $S_{ab} \equiv \nabla_a \xi_b + \nabla_b \xi_a$, and compute $\Box S_{ab}$. Observe that commuting covariant derivatives and using that by assumption $R_{ab} = 0$ and $\nabla^e R^f{}_{bea} = 0$ one has that

$$
\begin{aligned}
\Box \nabla_a \xi_b &= \nabla^e \nabla_e \nabla_a \xi_b = \nabla^e \nabla_a \nabla_e \xi_b + \nabla^e (R^f{}_{bea} \xi_f) \\
&= \nabla^e \nabla_a \nabla_e \xi_b + R^f{}_{bea} \nabla^e \xi_f, \\
&= \nabla_a \nabla^e \nabla_e \xi_b + R^f{}_e{}^e{}_a \nabla_f \xi_b + R^f{}_b{}^e{}_a \nabla_e \xi_f \\
&= \nabla_a \Box \xi_b + R^f{}_b{}^e{}_a \nabla_e \xi_f.
\end{aligned}
$$

So that

$$
\begin{aligned}
\Box S_{ab} &= R^e{}_a{}^f{}_b \nabla_f \xi_e + R^e{}_a{}^f{}_b \nabla_e \xi_f + \nabla_a \Box \xi_b + \nabla_b \Box \xi_a \\
&= R^e{}_a{}^f{}_b S_{ef} + \nabla_a \Box \xi_b + \nabla_b \Box \xi_a.
\end{aligned}
$$

Now, assume that one has a vector ξ^a satisfying the wave equation (7.1). One has then that

$$\Box S_{ab} - R^e{}_a{}^f{}_b S_{ef} = 0. \tag{7.2}$$

If initial data on an hypersurface $\mathcal{S}$ can be chosen such that

$$S_{ab} = 0, \quad \nabla_c S_{ab} = 0, \quad \text{on } \mathcal{S} \tag{7.3}$$

then, as a consequence of the homogeneity of equation (7.2), it follows that necessarily $S_{ab} = 0$ in the development of $\mathcal{S}$ so that ξ^a is, in fact, a Killing vector.

The conditions (7.3) are called the *Killing initial Data (KID) conditions*. They are conditions not only on ξ^a but also on the initial data $(\mathcal{S}, h_{ij}, K_{ij})$. In order to see this better, one can perform a $3+1$ split of the conditions. As a first step one writes $\xi^a = N n^a + N^a$ and $n_a N^a = 0$ where N and N^a denote the *lapse and shift of the Killing vector*. A computation then shows that the space-space components of the equation $\nabla_a \xi_b + \nabla_b \xi_a = 0$ imply (*exercise!*)

$$NK_{ij} + D_{(i}N_{j)} = 0.$$

Moreover, taking a time derivative of the above equation and using the ADM evolution equations one finds that (*exercise!*)

$$N^k D_k K_{ij} + D_i N^k K_{kj} + D_j N^k K_{ik} + D_i D_j N$$
$$= N\left(r_{ij} + KK_{ij} - 2K_{ik}K^k{}_j\right).$$

From the above expressions one can prove the following theorem.

Theorem 7.1. *Let (S, h_{ij}, K_{ij}) denote an initial data set for the vacuum Einstein field equations. If there exists a pair (N, N^i) such that*

$$NK_{ij} + D_{(i}N_{j)} = 0,$$
$$N^k D_k K_{ij} + D_i N^k K_{kj} + D_j N^k K_{ik} + D_i D_j N$$
$$= N(r_{ij} + KK_{ij} - 2K_{ik}K^k{}_j),$$

then the development of the initial data has a Killing vector.

The KID conditions are overdetermined. This is natural as not every spacetime admits a symmetry.

Remark. The KID conditions are closely related to the constraint equations and the ADM evolution equations — see e.g. the research survey by Fischer and Marsden [9].

8. Further Reading

If required, a good general introduction to modern General Relativity is given in:

— R. M. Wald, *General Relativity*. Chicago University Press (1984).

A number of introductory texts to various topics on mathematical General Relativity have appeared in recent years. Most notably:

— Y. Choquet-Bruhat, *General Relativity and Einstein's Equations*. Oxford University Press (2008).

— D. Christodoulou, *Mathematical Problems of General Relativity. I.* Zürich Lectures in Advanced Mathematics. European Mathematical Society (2008).
— A. D. Rendall, *Partial Differential Equations in General Relativity.* Oxford University Press (2008).
— H. Ringström, *The Cauchy Problem in General Relativity.* European Mathematical Society (2009).

References

[1] J. A. Valiente Kroon, *Conformal Methods in General Relativity.* Cambridge University Press (in preparation).
[2] R. M. Wald, *General Relativity.* The University of Chicago Press (1984).
[3] H. Stephani, D. Kramer, M. A. H. MacCallum, C. Hoenselaers and E. Herlt, *Exact Solutions of Einstein's Field Equations*, 2nd edn. Cambridge University Press (2003).
[4] J. B. Griffiths and J. Podolský, *Exact Space-Times in Einstein's General Relativity.* Cambridge University Press (2009).
[5] A. D. Rendall, *Partial Differential Equations in General Relativity.* Oxford University Press (2008).
[6] H. Friedrich and A. D. Rendall, The Cauchy problem for the Einstein equations. *Lect. Notes. Phys.* **540**, 127 (2000).
[7] T. W. Baumgarte and S. L. Shapiro, *Numerical Relativity: Solving Einstein's Equations on the Computer.* Cambridge University Press (2010).
[8] D. R. Brill and R. W. Lindquist, Interaction energy in geometrostatics. *Phys. Rev.* **131**, 471 (1963).
[9] A. E. Fischer and J. E. Marsden, The Einstein evolution equations as a first-order quasi-linear symmetric hyperbolic system, I. *Comm. Math. Phys.* **28**, 1 (1972).

Chapter 7

Value Distribution of Meromorphic Functions

Rod Halburd

Department of Mathematics, University College London
Gower Street, London, WC1E 6BT, UK
R.Halburd@ucl.ac.uk

Nevanlinna theory studies the distribution of values of meromorphic functions in the complex plane. We will derive Jensen's formula and use it to prove Nevanlinna's First Main Theorem, which quantifies a balance between how often in a large disc a function takes a given value and how close the function stays to that value on the boundary of the disc. The Lemma on the Logarithmic Derivative is a fundamental estimate in Nevanlinna theory. Here it is used to show that meromorphic solutions of certain differential equations have infinitely many poles. It is also used to derive Nevanlinna's Second Main Theorem, which is a profound generalisation and quantification of Picard's Theorem, which says that an entire function that misses two values must be a constant.

1. Introduction

A function that is analytic at every point of $\mathbb{C}$ is said to be *entire*. A function that is analytic at every point of $\mathbb{C}$ except at points where it has poles is said to be *meromorphic*. Many commonly used functions are meromorphic such as the trigonometric and hyperbolic functions, elliptic functions, polynomials and rational functions. In this chapter we will study the value distribution of meromorphic functions known as Nevanlinna theory.

If a function f is analytic at $z = z_0$ and

$$f(z) = a + \alpha(z - z_0)^p + O\left((z - z_0)^{p+1}\right)$$

near z_0, where $\alpha \neq 0$, then we call z_0 an a-point of f of multiplicity p. If $a = \infty$ then an a-point of multiplicity p means a pole of multiplicity p. The large-scale value distribution of a polynomial is determined by its degree. The Fundamental Theorem of Algebra says that given any $a \in \mathbb{C}$, a polynomial of degree d takes the value a exactly d times, counting multiplicities, in the complex plane. For example, the degree three polynomial $f(z) = z^3 - 6z^2 + 12z$ takes the value 0 at $z = 0$, $z = 3 + i\sqrt{3}$ and $z = 3 - i\sqrt{3}$, each with multiplicity one. Since we can write $f(z) = 8 + (z - 2)^3$, we see that f has exactly one 8-point at $z = 2$ of multiplicity three, so it has three 8-points counting multiplicities. Notice that the degree d of a polynomial is encoded in the fact that $|f(z)| \sim Kr^d$ as $r = |z| \to \infty$.

For an entire function f the number of a-points in a large disc centred at the origin has a bound in terms of the maximum modulus of f on the disc (which is the same as the maximum modulus on the bounding circle). For a genuinely meromorphic function f, a large amount of information about value distribution is encoded in the Nevanlinna characteristic $T(r, f)$, which is a real-valued function of r containing information about how large f is on the circle $|z| = r$ and the poles in the disc $|z| \leq r$.

In Section 2 we will derive Jensen's formula and use it to introduce the main tools of Nevanlinna theory such as the Nevanlinna characteristic, counting and proximity functions. Nevanlinna's First Main Theorem is effectively a rewriting of Jensen's formula as a certain balance between the number of a-points of meromorphic function (counting multiplicities) in a large disc and how close the function stays to a on the boundary of the disc. Section 3 contains a fundamental estimate, the Lemma on the Logarithmic Derivative, which says that on most large circles, a certain average of the logarithmic derivative f'/f is small compared to the Nevanlinna characteristic of f. Applications to differential equations are considered.

Section 4 contains the most important application of the Lemma on the Logarithmic Derivative, namely Nevanlinna's Second Main Theorem and the resulting defect relations. Some of the profound consequences of these results include Picard's theorem, which says that a meromorphic function missing three values in the extended complex plane (i.e. one of the values could be infinity) must be a constant. In fact, a meromorphic function that takes three values only a finite number of times must be rational. Another implication is that, in a certain sense, there can be at most a countable number of values that a non-constant meromorphic function can take fewer times than usual. Finally, a non-rational meromorphic function can have at most four totally ramified point, i.e. there can be at most four values a in the extended complex plane such that each a-point occurs with multiplicity greater than one.

Finally, in Section 5 we indicate some further reading.

2. Nevanlinna's First Main Theorem

2.1. *Jensen's formula*

Suppose that f is analytic and nowhere vanishing in the closed disc $D = \{z : |z| \leq r\}$. Then $\log f(z)$ is analytic in D. In particular note that $\log f(z)$ is not branched although we do have to choose a particular branch to have a well-defined function since its imaginary part is defined up to the addition of integer multiples of 2π. Taking the real part of Cauchy's integral formula for $\log f$,

$$\log f(0) = \frac{1}{2\pi i} \int_{|z|=r} \frac{\log f(z)}{z} dz,$$

gives

$$\log |f(0)| = \frac{1}{2\pi} \int_0^{2\pi} \log \left| f(re^{i\theta}) \right| d\theta. \tag{2.1}$$

Now let f be any meromorphic function which is not identically zero. Then f has finitely many zeros $a_1, \ldots, a_m$ and poles $b_1, \ldots, b_n$ in $D\backslash\{0\}$. We will consider the case in which there are no zeros or poles on the boundary circle $|z| = r$. Furthermore, suppose that the Laurent series expansion of f at $z = 0$ has the form

$$f(z) = \alpha z^\nu + \cdots,$$

where $\alpha \neq 0$. Define $\text{ilc}_0 f = \alpha$ and $\text{ord}_0 f = \nu$. With these definitions, the function

$$g(z) := z^{-\text{ord}_0 f} \frac{\prod B(a_j, z)}{\prod B(b_k, z)} f(z),$$

where

$$B(a, z) = \frac{r^2 - \bar{a}z}{r(z - a)},$$

has no zeros or poles in D and $|B(a,z)| = 1$ on the circle $\partial D = \{z : |z| = r\}$. Now we write equation (2.1) for g:

$$
\begin{aligned}
\log|\text{ilc}_0 f| = \frac{1}{2\pi} \int_0^{2\pi} \log|f(re^{i\theta})|\, d\theta \\
+ \sum \log \frac{r}{|b_k|} - \sum \log \frac{r}{|a_j|} - (\text{ord}_0 f) \log r.
\end{aligned}
\tag{2.2}
$$

Let $n(r,f)$ be the number of poles of f (counting multiplicities) in $|z| \le r$. Then we define the integrated counting function $N(r,f)$ by

$$
N(r,f) := \int_0^r \frac{n(t,f) - n(0,f)}{t}\, dt + n(0,f) \log r.
$$

It follows (exercise) that

$$
N(r,f) = \sum_{k=1}^n \log \frac{r}{|b_k|} + n(0,f) \log r.
\tag{2.3}
$$

Since $\text{ord}_0 f = n(0,1/f) - n(0,f)$, equation (2.2) becomes Jensen's formula,

$$
\log|\text{ilc}_0 f| = \frac{1}{2\pi} \int_0^{2\pi} \log|f(re^{i\theta})|\, d\theta + N(r,f) - N(r,1/f).
\tag{2.4}
$$

The starting point for Nevanlinna theory is the realisation that Jensen's formula can be written in a very symmetric way in which the zeros and the poles of f play an equal role. For any $x > 0$, define

$$
\log^+ x := \max(\log x, 0).
$$

Then

$$
\log x = \log^+ x - \log^+(x^{-1}).
$$

Jensen's formula (2.4) can now be written as

$$
\begin{aligned}
&\frac{1}{2\pi} \int_0^{2\pi} \log^+ |f(re^{i\theta})|\, d\theta + N(r,f) \\
&= \frac{1}{2\pi} \int_0^{2\pi} \log^+ \left| \frac{1}{f(re^{i\theta})} \right| d\theta + N(r,1/f) + \log|\text{ilc}_0 f|.
\end{aligned}
\tag{2.5}
$$

2.2. The Nevanlinna characteristic

Motivated by equation (2.5), we define the *proximity function* $m(r, f)$ by

$$m(r, f) = \frac{1}{2\pi} \int_0^{2\pi} \log^+ |f(re^{i\theta})| \, d\theta.$$

The proximity function is large if f is large on some non-negligible part of the circle $|z| = r$. Since $m(r, f)$ is the average of $\log^+ |f|$ on $|z| = r$, as opposed to $\log |f|$, there are no negative contributions when $|f|$ is small. We should think of $m(r, f)$ as a measure of how close f is to infinity on $|z| = r$.

The *Nevanlinna characteristic*

$$T(r, f) = m(r, f) + N(r, f)$$

measures "the affinity" of f for infinity. Similarly, for all $a \in \mathbb{C}$,

$$T\left(r, \frac{1}{f-a}\right) = m\left(r, \frac{1}{f-a}\right) + N\left(r, \frac{1}{f-a}\right)$$

measures the "affinity" of f for the value a. $T(r, 1/(f-a))$ is large if f is close to a on a significant part of the circle $|z| = r$ or if it has many a-points in the disc $|z| \leq r$.

Jensen's formula becomes

$$T(r, f) = T(r, 1/f) + \log |\text{ilc}_0 \, f|. \tag{2.6}$$

We will see that if f is not constant then $T(r, f) \to \infty$ as $r \to \infty$. So equation (2.6) says that asymptotically, the affinity of f for ∞ is approximately the same as the affinity of f for 0. Nevanlinna's First Main Theorem below extends this to all values $a \in \mathbb{C}$.

Lemma 2.1 (Elementary properties of $\log^+$). *Let $a_1, \ldots, a_q$ be positive real numbers. Then*

$$\log^+ \left(\prod_{j=1}^q a_j \right) \leq \sum_{j=1}^q \log^+ a_j,$$

$$\log^+ \left(\sum_{j=1}^q a_j \right) \leq \log^+ \left(q \max_{1 \leq j \leq q} a_j \right) \leq \sum_{j=1}^q \log^+ a_j + \log q,$$

$$\log a = \log^+ a - \log^+(1/a),$$

$$|\log a| = \log^+ a + \log^+(1/a),$$

$$\log^+ a \leq \log^+ b, \quad \forall a \leq b.$$

Lemma 2.2 (Elementary properties of the Nevanlinna functions).
Let $f_1, \ldots, f_q$ be non-zero meromorphic functions. Then

$$n\left(r, \sum_{j=1}^{q} f_j\right) \leq \sum_{j=1}^{q} n(r, f_j), \qquad n\left(r, \prod_{j=1}^{q} f_j\right) \leq \sum_{j=1}^{q} n(r, f_j),$$

$$N\left(r, \sum_{j=1}^{q} f_j\right) \leq \sum_{j=1}^{q} N(r, f_j), \qquad N\left(r, \prod_{j=1}^{q} f_j\right) \leq \sum_{j=1}^{q} N(r, f_j),$$

$$m\left(r, \sum_{j=1}^{q} f_j\right) \leq \sum_{j=1}^{q} m(r, f_j) + \log q, \quad m\left(r, \prod_{j=1}^{q} f_j\right) \leq \sum_{j=1}^{q} m(r, f_j),$$

$$T\left(r, \sum_{j=1}^{q} f_j\right) \leq \sum_{j=1}^{q} T(r, f_j) + \log q, \quad T\left(r, \prod_{j=1}^{q} f_j\right) \leq \sum_{j=1}^{q} T(r, f_j).$$

Proof. Exercise. □

We now have the simple tools needed to prove Nevanlinna's First Main Theorem.

Theorem 2.3 (Nevanlinna's First Main Theorem). *For any $a \in \mathbb{C}$ and any meromorphic function $f \not\equiv a$, we have*

$$T\left(r, \frac{1}{f-a}\right) = T(r, f) + O(1), \quad r \to \infty.$$

Proof. Now

$$T(r, f-a) \leq T(r, f) + T(r, a) + \log 2.$$

Similarly

$$T(r, f) = T(r, (f-a) + a) \leq T(r, f-a) + T(r, a) + \log 2.$$

Hence

$$|T(r, f-a) - T(r, f)| \leq T(r, a) + \log 2 = \log^+ |a| + \log 2.$$

Therefore

$$\left| T(r,f) - T\left(r, \frac{1}{f-a}\right) \right|$$

$$\leq |T(r,f) - T(r, f-a)| + \left| T(r, f-a) - T\left(r, \frac{1}{f-a}\right) \right|$$

$$\leq \log^+ |a| + \log 2 + \log^+ |\mathrm{ilc}_a f|. \qquad \square$$

Example. We begin by writing down the First Main Theorem for the function $\exp(z)$ with $a = 1$.

$$m\left(r, \frac{1}{e^z - 1}\right) + N\left(r, \frac{1}{e^z - 1}\right) = m(r, e^z) + N(r, e^z) + O(1).$$

Now $\exp(z) = 1$ if and only if $z = 2k\pi i$ for $k \in \mathbb{Z}$. Furthermore, each 1-point has multiplicity one. Therefore $n(r, 1/(e^z - 1)) = 1 + 2[r/(2\pi)]$, where $[x]$ is the integer part of x. So $(r/\pi) - 1 \leq n(r, 1/(e^z - 1)) \leq (r/\pi) + 1$ and $N(r, 1/(e^z - 1)) = (r/\pi) + O(\log r)$. The fact that as we go from considering 1-points to poles, the N term drops from something with significant growth to 0 means that the $m(r, e^z)$ must be large. This means that f must be "close to infinity" on a significant part of the circle $|z| = r$, which is indeed the case in the right half-plane. Nevanlinna's First Main Theorem is really saying that since the function $\exp(z)$ is never zero it must stay near this values on large parts of the circle $|z| = r$ for $r \gg 1$. The same is true for the value 0.

The following theorem gives another useful representation of $T(r, f)$ due to Cartan.

Theorem 2.4. *Let f be a non-constant meromorphic function, with $f(0)$ finite and non-zero. Then*

$$T(r,f) = \frac{1}{2\pi} \int_0^{2\pi} N\left(r, \frac{1}{f(z) - e^{i\theta}}\right) d\theta + \log^+ |f_0|.$$

Proof. Applying Jensen's formula (2.4) with $r = 1$ to $f(z) = z - a$, where $a \neq 0$, gives

$$\frac{1}{2\pi} \int_0^{2\pi} \log |e^{i\theta} - a| d\theta = \log |a| + N\left(1, \frac{1}{z-a}\right) = \log^+ |a|. \qquad (2.7)$$

Now applying Jensen's formula (2.4) to $f(z) - e^{i\theta}$ gives

$$\log |f(0) - e^{i\theta}| = \frac{1}{2\pi} \int_0^{2\pi} \log \left| f(re^{i\phi}) - e^{i\theta} \right| d\phi + N(r, f)$$

$$- N\left(r, \frac{1}{f(z) - e^{i\theta}}\right).$$

On dividing by 2π, integrating with respect to θ and reversing the order of integration in the resulting double integral, we obtain

$$\frac{1}{2\pi} \int_0^{2\pi} \log |f(0) - e^{i\theta}| d\theta$$

$$= \frac{1}{2\pi} \int_0^{2\pi} \frac{1}{2\pi} \int_0^{2\pi} \log |f(re^{i\phi}) - e^{i\theta}| \, d\theta \, d\phi \qquad (2.8)$$

$$+ N(r, f) - \frac{1}{2\pi} \int_0^{2\pi} N\left(r, \frac{1}{f(z) - e^{i\theta}}\right) d\theta.$$

Finally we apply equation (2.7) twice in equation (2.8), once with $a = f(0)$ and once with $a = f(re^{i\phi})$, to give

$$\log^+ |f_0|$$

$$= \frac{1}{2\pi} \int_0^{2\pi} \log^+ |f(re^{i\theta})| \, d\theta + N(r, f) - \frac{1}{2\pi} \int_0^{2\pi} N\left(r, \frac{1}{f(z) - e^{i\theta}}\right) d\theta$$

$$= m(r, f) + N(r, f) - \frac{1}{2\pi} \int_0^{2\pi} N\left(r, \frac{1}{f(z) - e^{i\theta}}\right) d\theta,$$

which completes the proof. $\square$

Some important properties of $T(r, f)$ follow immediately from this theorem.

Corollary 2.5. *For any meromorphic function f, the Nevanlinna characteristic $T(r, f)$ is a continuous non-decreasing function of r. Furthermore, $T(r.f)$ is a convex function of $\log r$.*

Lemma 2.6. *For all $K > 1$,*

$$n(r, f) \le \frac{1}{\log K} N(Kr, f) \qquad (2.9)$$

Proof. From the definition of N we have

$$N(Kr, f) = \int_0^{Kr} \frac{n(t, f) - n(0, f)}{t} dt + n(0, f) \log Kr$$

$$\ge \int_r^{Kr} \frac{n(t, f) - n(0, f)}{t} dt + n(0, f) \log Kr$$

$$= \int_r^{Kr} \frac{n(t, f)}{t} dt + n(0, f) \log r$$

$$\ge n(r, f) \int_r^{Kr} \frac{1}{t} dt = n(r, f) \log K.$$ $\square$

Theorem 2.7. *Let f be a non-constant entire function. Let $r > 0$ be sufficiently large that $M(r, f) := \max_{|z|=r} |f(z)| \geq 1$. Then for all finite $R > r$ we have*

$$T(r, f) \leq \log M(r, f) \leq \frac{R+r}{R-r} T(R, f).$$

Proof. See exercises. □

2.3. Rational functions in Nevanlinna theory

Recall that a rational function is a ratio of polynomials. If $f(z) = P(z)/Q(z)$, where P and Q are polynomials with no common factors, then the degree of f is defined to be the maximum of the degrees of P and Q. Rational functions are the simplest kinds of meromorphic functions after constants.

Theorem 2.8. *A meromorphic function f is rational if and only if*

$$T(r, f) = O(\log r).$$

Proof. The fact that $T(r, f) = O(\log r)$ implies that f is rational follows from the next theorem. The reverse implication is left as an exercise. □

Theorem 2.9. *A meromorphic function f is rational if*

$$\liminf_{r \to \infty} \frac{T(r, f)}{\log r}$$

is finite.

Proof. If $\liminf_{r \to \infty} \frac{T(r,f)}{\log r}$ is finite then there exist a constant C and a sequence $s_j \to \infty$ such that $N(s_j, f) \leq T(s_j, f) \leq C \log s_j$ for all j. Let $r_j = \sqrt{s_j}$; then using the inequality (2.9) with $K = r_j$, we have for $r > 1$

$$n(r_j, f) \leq N(r_j^2, f)/\log r_j \leq 2C.$$

Hence f has only finitely many poles. So there is a polynomial $p(z)$ such that $F(z) := p(z)f(z)$ is an entire function. Moreover, we know from Theorem 2.8 that $T(r, p) = O(\log r)$. Hence, for all j, $T(s_j, F) \leq \lambda \log s_j$ for some constant λ and so from Theorem 2.7, $\log M(s_j, F) \leq \kappa \log s_j$ for some κ. Therefore $M(s_j, F) \leq s_j^\kappa$. Hence by (the extended version of) Liouville's Theorem, F must be a polynomial. □

2.4. *Exercises*

(1) Derive equation (2.3).

(2) Finish the proof of Theorem 2.8.

(3) Show that for any meromorphic functions f, g and h,

$$T(r, fg + gh + hf) \leq T(r, f) + T(r, g) + T(r, h) + \log 3.$$

(4) For any meromorphic function f and any $q \in \mathbb{C}\backslash\{0\}$, define $g(z) = f(qz)$. Show that $T(r, g) = T(|q|r, f) - n(0, f) \log |q|$.

(5) Let $a_0, \ldots, a_d$ be rational functions and let f be a meromorphic function. Show that

$$T(r, a_0 + a_1 f + \cdots + a_d f^d) \leq dT(r, f) + O(\log r).$$

(6) Let f be meromorphic on the closed disc D of radius $R > 0$, $f \not\equiv 0$. Denote the zeros and poles of f in D by a_μ and b_ν respectively. Prove the Poisson–Jensen formula:

$$\log |f(z)| = \frac{1}{2\pi} \int_0^{2\pi} \log |f(Re^{i\phi})| \Re \left(\frac{Re^{i\phi} + z}{Re^{i\phi} - z} \right) d\phi$$

$$- \sum_{|a_\mu| < R} \log \left| \frac{R^2 - \bar{a}_\mu z}{R(z - a_\mu)} \right| + \sum_{|b_\mu| < R} \log \left| \frac{R^2 - \bar{b}_\mu z}{R(z - b_\mu)} \right|.$$

Hint: Fix z and let $F(w) = f(z)$, where $w = (Re^{i\phi} - z)/(R - \bar{z}e^{i\phi})$. Note that $w = 0$ corresponds to $z = Re^{i\phi}$. For $|z| < R$, $|w| = 1$.

(7) Prove Theorem 2.7. Hint: use the Poisson–Jensen formula from problem (2.6).

3. Logarithmic Derivative Estimates and Applications

3.1. *Logarithmic derivatives*

For any non-zero rational function f, the logarithmic derivative $f'(z)/f(z)$ tends to zero as $z \to \infty$. The *Lemma on the Logarithmic Derivative* is a generalisation of this result to meromorphic functions which states that $m(r, f) = o(T(r, f))$ as $r \to \infty$ outside of some (possibly empty) set E of finite linear (Lebesgue) measure. In other words $\int_E dr < \infty$. It is this estimate of the size of logarithmic derivatives that is usually more accurate and useful than any estimate on the size of the derivative f'. We will not prove the Lemma on the Logarithmic Derivative in these notes but we will use it to prove some global results about meromorphic solutions of some differential equations.

The most important consequence of the Lemma on the Logarithmic Derivative is Nevanlinna's Second Main Theorem, which is the central result of Nevanlinna theory. It is a vast generalisation and quantification of Picard's Theorem. It bounds the growth of a meromorphic function (as measured by the characteristic $T(r, f)$) by a measure of how many times f takes a fixed, finite collection of values in the extended complex plane. We will prove this result in Section 4. There has been renewed interest recently in finding better estimates of the error terms in these theorems.

3.2. *Fundamental estimates*

We will denote by $S(r, f)$ any real-valued function $s(r)$ satisfying $s(r) = o(T(r, f))$ outside of a possible exceptional set of finite linear measure. We use this notation in the same way that one uses the big-O and little-o notation in asymptotics in the sense that the symbol $S(r, f)$ can appear several times throughout a calculation and represent different terms. The notation just reflects the fact that these terms are small outside some possible exceptional set.

Theorem 3.1 (Lemma on the Logarithmic Derivative). *Let f be a non-constant meromorphic function. Then*

$$m(r, f'/f) = S(r, f).$$

Furthermore, if f has finite order then

$$m(r, f'/f) = O(\log r).$$

The order of meromorphic functions. The *order* of a meromorphic function f is

$$\sigma(f) = \limsup_{r \to \infty} \frac{\log T(r, f)}{\log r}.$$

Note that there is no exceptional set in the claim about finite-order functions in Theorem 3.1. There is a lot of technical detail that enters many proofs in Nevanlinna theory to deal with these exceptional sets.

Corollary 3.2. *For any non-constant meromorphic function f and positive integer k,*

$$m(r, f^{(k)}/f) = S(r, f) \quad and \quad T(r, f^{(k)}) \le (k+1)T(r, f) + S(r, f). \quad (3.1)$$

Proof. We prove the first statement by induction and, in so doing, we will also prove the second statement. When $k = 1$, the first statement is just the Lemma on the Logarithmic Derivative. Assume that we have proved

$$m(r, f^{(l)}/f) = S(r, f),$$

for some $l \ge 1$. Then

$$m(r, f^{(l)}) \le m(r, f) + m(r, f^{(l)}/f) = m(r, f) + S(r, f).$$

The meromorphic function f has a pole of order p at some point z_0 if and only if $f^{(l)}$ has a pole of order $l + p \ge l + 1$ at $z = z_0$. It follows that $n(r, f^{(l)}) \le (l+1)n(r, f)$. Therefore

$$N(r, f^{(l)}) \le (l+1)N(r, f).$$

So

$$T(r, f^{(l)}) \le (l+1)T(r, f) + S(r, f).$$

This is the second claim in the statement of the lemma in the case $k = l$. Hence

$$m(r, f^{(l+1)}/f^{(l)}) = S(r, f^{(l)}) = S(r, f).$$

Finally

$$m(r, f^{(l+1)}/f) \le m(r, f^{(l+1)}/f^{(l)}) + m(r, f^{(l)}/f)$$
$$= S(r, f) + S(r, f) = S(r, f). \qquad \square$$

3.3. *Applications to differential equations*

We have already developed enough tools to be able to conclude some remarkable global results about solutions of differential equations.

Theorem 3.3. *Every meromorphic solution of the first Painlevé equation,*

$$y'' - 6y^2 + z, \quad (3.2)$$

has infinitely many poles.

We remark that it can be shown that any solution of equation (3.2) can be analytically extended to a meromorphic solution.

Proof of Theorem 3.3. Assume first that y is transcendental and write equation (3.2) as

$$y^2 = 6^{-1}\left(y\frac{y''}{y} - z\right).$$

An obvious property of the proximity function m is $m(r, y^2) = 2m(r, y)$. Hence

$$2m(r, y) = m(r, y^2) = m\left(r, 6^{-1}\left(y\frac{y''}{y} - z\right)\right)$$

$$\leq m(r, 6^{-1}) + m\left(r, y\frac{y''}{y} - z\right)$$

$$\leq m\left(r, y\frac{y''}{y}\right) + m(r, z) + \log 2$$

$$\leq m(r, y) + m\left(r, \frac{y''}{y}\right) + O(\log r)$$

$$= m(r, y) + S(r, y) + O(\log r).$$

So $m(r, y) = S(r, y) + O(\log r)$. Suppose that y has only finitely many poles. Then $N(r, y) = O(\log r)$. Therefore

$$T(r, y) = m(r, y) + N(r, y) = S(r, y) + O(\log r).$$

Recall from Theorem 2.9 that if y is a transcendental meromorphic function then $\log r = o(T(r, y))$. Therefore our solution y satisfies $T(r, y) = S(r, y)$, which means that $T(r, y) = o(T(r, y))$ as $r \to \infty$ outside of some possible exceptional set E of finite linear measure, which is clearly a contradiction.

Now assume that y is rational. Clearly there are no constant solutions of equation (3.2). Therefore, there exist an integer μ and a non-zero constant c such that $y = cz^\mu + \cdots$ as $z \to \infty$. Substituting this into equation (3.2) and keeping only the possible leading-order terms gives

$$\mu(\mu - 1)cz^{\mu-2} + \cdots = 6c^2z^{2\mu} + \cdots + z.$$

So if $\mu \neq 0$ or 1 (which guarantees that the coefficient of $z^{\mu-2}$ does not vanish), then at least two of the powers of z appearing above (i.e. $\mu - 2$, 2μ, 1) must be the same. If $\mu - 2$ and 2μ are the leading powers (i.e. the largest powers), then $\mu - 2 = 2\mu$ and so $\mu = -2$. But then z would be the leading term, which is a contradiction. Similarly, if $\mu - 2$ and 1 are the leading powers, then $\mu = 3$ but then $2\mu = 6$ would be the leading power. Finally we are left with $2\mu = 1$. But then $\mu = 1/2$, which is not an integer. If $\mu = 0$ or 1 then there is only one leading term (proportional to z or z^2 respectively as $z \to \infty$) with nothing to balance it. So there are no rational solutions of equation (3.2). $\square$

The proof of the following result uses only elementary estimates but we will not include it here (see, e.g., [6]).

Theorem 3.4 (Valiron-Mohon'ko). *Let*

$$R(z, f(z)) := \frac{a_0(z) + a_1(z)f(z) + \cdots + a_p(z)f^p(z)}{b_0(z) + b_1(z)f(z) + \cdots + b_q(z)f^q(z)}$$

be a rational function of f of degree $d = \max(p, q)$ with coefficients a_i and b_j satisfying

$$T(r, a_i) = S(r, f) \quad and \quad T(r, b_j) = S(r, f).$$

Then

$$T(r, R(z, f(z))) = d\,T(r, f) + S(r, f).$$

Theorem 3.5 (Malmquist's Theorem). *Let f be a meromorphic solution of the equation*

$$f'(z) = R(z, f(z)) := \frac{a_0(z) + a_1(z)f(z) + \cdots + a_p(z)f^p(z)}{b_0(z) + b_1(z)f(z) + \cdots + b_q(z)f^q(z)}, \tag{3.3}$$

where the coefficients a_i and b_j satisfy

$$T(r, a_i) = S(r, f) \quad and \quad T(r, b_j) = S(r, f).$$

Then equation (3.3) has the form

$$f'(z) = a_0(z) + a_1(z)f(z) + a_2(z)f^2(z). \tag{3.4}$$

If $a_2 \equiv 0$ then equation (3.4) is a linear equation. If $a_2 \not\equiv 0$ then equation (3.4) is a Riccati equation and its general solution is given in terms of the general solution of a second-order linear equation.

Proof. Using Theorem 3.4 and the second part of Theorem 2.4 with $k = 1$, we have

$$d\,T(r, f) + S(r, f) = T(r, R(z, f(z))) = T(r, f') \leq 2T(r, f) + S(r, f).$$

Hence $d \leq 2$. It remains to show that R is a polynomial. Write $R = P/Q$, where

$$P(z, f) = a_0(z) + a_1(z)f + a_2(z)f^2 \quad \text{and} \quad Q(z, f) = b_0(z) + b_1(z)f + b_2(z)f^2,$$

are relatively prime (i.e. have no common factors) as polynomials in f. Without loss of generality (on shifting f by a constant if necessary: $f \mapsto f + k$) we assume that $a_0(z) \not\equiv 0$. Now the function $g := 1/f$ satisfies the equation

$$g' = -\frac{g^2(a_0 g^2 + a_1 g + a_2)}{b_0 g^2 + b_1 g + b_2} = \widetilde{R}(z, g) = \frac{\widetilde{P}(z, g)}{\widetilde{Q}(z, g)}, \tag{3.5}$$

where $\widetilde{P}(z, g) = -g^2(a_0 g^2 + a_1 g + a_2) = -g^4 P(z, 1/g)$ and $\widetilde{Q}(z, g) = g^2 Q(z, 1/g)$. From the First Main Theorem we have $T(r, g) = T(r, f) + O(1)$. Hence $T(r, a_i) = S(r, g)$ and $T(r, b_j) = S(r, g)$. Also, equation (3.5) has the same general form as equation (3.3), so $\widetilde{R}$ has degree at most 2. Therefore two of the roots (counting multiplicities) of the quartic polynomial $\widetilde{P} = -g^4 P(z, 1/g)$ must be shared by $\widetilde{Q}(z, g) = z^2 Q(z, 1/g)$. Recall that P and Q are relatively prime and 0 is not a root of P (since $a_0 \not\equiv 0$). So g^2 must divide $\widetilde{Q}(z, 1/g)$. Hence $b_1 = b_2 = 0$ and R is a polynomial of degree at most 2 as required. $\qquad\square$

3.4. Clunie-type lemmas

In the proof of Theorem 3.3, we showed that any transcendental meromorphic solution y of the first Painlevé equation has infinitely many poles by showing that $m(r, y) = S(r, y)$ by a direct use of the Lemma on the Logarithmic Derivative. Here we provide two important lemmas that can be used to derive similar conclusions for many differential equations of arbitrary order.

Theorem 3.6 (A. Mohon'ko and V. Mohon'ko). *Let f be a transcendental meromorphic solution of*

$$P(z; f, f', \ldots, f^{(n)}) = 0, \tag{3.6}$$

where P is a non-zero polynomial in its arguments. If the constant $a \in \mathbb{C}$ does not solve equation (3.6), then

$$m\left(r, \frac{1}{f - a}\right) = S(r, f).$$

Proof. Without loss of generality (by considering the equation solved by $\tilde{f} := f - a$ if necessary), we take $a = 0$. Now

$$0 = P(z; f, f', \ldots, f^{(n)}) = p(z) + Q(z; f, f', \ldots, f^{(n)}),$$

where $p(z) = P(z; 0, \ldots, 0)$ and

$$Q(z; f, f', \ldots, f^{(n)}) = \sum_{\lambda = (\lambda_0, \ldots, \lambda_n) \in \Lambda} q_\lambda f^{\lambda_0} (f')^{\lambda_1} \cdots (f^{(n)})^{\lambda_n},$$

where Λ is a finite set and each $\lambda = (\lambda_0, \ldots, \lambda_n) \in \Lambda$ satisfies $|\lambda| = \lambda_0 + \cdots + \lambda_n \geq 1$.

In calculating $m(r, 1/f)$ we need only consider those z such that $|z| = r$ where $|f(z)| < 1$. For each such z and each $(\lambda_0, \ldots, \lambda_n) \in \Lambda$ we have

$$\frac{1}{|f|} \leq \frac{1}{|f|^{\lambda_0 + \cdots + \lambda_n}}.$$

Therefore,

$$\left| \frac{Q(z; f, f', \ldots, f^{(n)})}{f} \right| \leq \sum_{\lambda = (\lambda_0, \ldots, \lambda_n) \in \Lambda} \left| q_\lambda \frac{f^{\lambda_0} (f')^{\lambda_1} \cdots (f^{(n)})^{\lambda_n}}{f} \right|$$

$$\leq \sum_{\lambda = (\lambda_0, \ldots, \lambda_n) \in \Lambda} |q_\lambda| \left| \frac{f'}{f} \right|^{\lambda_1} \cdots \left| \frac{f^{(n)}}{f} \right|^{\lambda_n}.$$

So by the Lemma on the Logarithmic Derivative (Theorem 3.1), $m(r, Q/f) = S(r, f)$. Therefore

$$m(r, 1/f) \leq m(r, Q/f) + m(r, 1/Q) = m(r, Q/f) + m(r, 1/p) = S(r, f).$$
$\square$

Theorem 3.7 (Clunie's Lemma). *Let f be a transcendental solution of*

$$f^N P(z, f) = Q(z, f), \tag{3.7}$$

where P and Q are differential polynomials in f with coefficients that are $S(r, f)$. If the total degree of Q is no greater than N, then

$$m(r, P(z, f)) = S(r, f).$$

Proof. Now

$$m(r, P) = \frac{1}{2\pi} \int_{z \in \Theta_+} \log^+ |P| \, d\theta + \frac{1}{2\pi} \int_{z \in \Theta_-} \log^+ |P| \, d\theta,$$

where

$$\Theta_+ = \{\theta \in [0, 2\pi) : |f(re^{i\theta})| \geq 1\},$$
$$\Theta_- = \{\theta \in [0, 2\pi) : |f(re^{i\theta})| < 1\}.$$

Let

$$P(z, f) = \sum_{\lambda = (\lambda_0, \ldots, \lambda_n) \in \Lambda} P_\lambda(z; f, f', \ldots, f^{(n)})$$

$$= \sum_{\lambda = (\lambda_0, \ldots, \lambda_n) \in \Lambda} p_\lambda f^{\lambda_0} (f')^{\lambda_1} \cdots (f^{(n)})^{\lambda_n},$$

for some finite index set Λ. For $z = re^{i\theta}$ with $\theta \in \Theta_-$ we have

$$|P_\lambda| \leq |p_\lambda| \left| \frac{f'}{f} \right|^{\lambda_1} \cdots \left| \frac{f^{(n)}}{f} \right|^{\lambda_n}.$$

Hence

$$\frac{1}{2\pi} \int_{z \in \Theta_-} \log^+ |P_\lambda| \, d\theta \leq m(r, p_\lambda) + \sum_{j=1}^n \lambda_j m\left(r, \frac{f^{(j)}}{f}\right) = S(r, f).$$

Now let

$$Q(z, f) = \sum_{\lambda = (\lambda_0, \ldots, \lambda_n) \in \tilde{\Lambda}} Q_\lambda(z; f, f', \ldots, f^{(n)})$$

$$= \sum_{\lambda = (\lambda_0, \ldots, \lambda_n) \in \tilde{\Lambda}} q_\lambda f^{\lambda_0} (f')^{\lambda_1} \cdots (f^{(n)})^{\lambda_n},$$

where for each $\lambda = (\lambda_0, \ldots, \lambda_n) \in \tilde{\Lambda}$ we have $|\lambda| = \lambda_0 + \lambda_1 + \cdots + \lambda_n \leq N$. It follows that for all $z = re^{i\theta}$ with $\theta \in \Theta_+$,

$$|P(z, f)| \leq \left| \frac{1}{f^n} \sum_{\lambda = (\lambda_0, \ldots, \lambda_n) \in \tilde{\Lambda}} q_\lambda f^{\lambda_0} (f')^{\lambda_1} \cdots (f^{(n)})^{\lambda_n} \right|$$

$$\leq \sum_{\lambda = (\lambda_0, \ldots, \lambda_n) \in \tilde{\Lambda}} |q_\lambda| \left| \frac{f'}{f} \right|^{\lambda_1} \cdots \left| \frac{f^{(n)}}{f} \right|^{\lambda_n}.$$

So

$$\frac{1}{2\pi} \int_{z \in \Theta_+} \log^+ |P| \, d\theta = S(r, f)$$

and therefore $m(r, P) = S(r, f)$. $\square$

3.5. *Exercises*

(1) Show that any transcendental meromorphic solution of the second Painlevé equation,

$$y'' = 2y^3 + zy + \alpha,$$

where $\alpha \neq 0$ is a constant, takes every value in the extended complex plane (i.e. including infinity) an infinite number of times.

(2) Show that if a transcendental meromorphic function f satisfies an equation of the form

$$f'(z)^2 = P(z, f(z)),$$

where $P(z, f)$ is a polynomial in its arguments, then the degree of P as a polynomial in y is at most 4.

(3) Find the order of $\cosh z$.

(4) The following is a difference version of the Lemma on the Logarithmic Derivative for finite-order meromorphic functions:

$$m\left(r, \frac{f(z+c)}{f(z)}\right) = o(T(r, f(z))),$$

as $r \to \infty$ outside a possible exceptional set E of finite logarithmic measure:

$$\int_{E \cap [1, \infty)} \frac{dr}{r} < \infty.$$

Use this lemma to prove a natural difference analogue of Theorem 3.6 for finite-order meromorphic solutions of difference equations. Hence show that any transcendental finite-order meromorphic solution of the difference equation

$$y(z+1) + y(z-1) = \frac{zy(z) + \gamma}{1 - y(z)^2},$$

where γ is a constant, takes every $a \in \mathbb{C}$ an infinite number of times.

4. The Second Main Theorem

4.1. *The main result*

Another way of writing the counting function $n(r, f)$ is

$$n(r, f) = \sum_{\zeta \in D_r} \max\{\mathrm{ord}_\zeta(1/f), 0\}.$$

This sum makes sense as it has only a finite number of non-zero terms. Consider

$$n_{\mathrm{ram}}(r, f) := 2n(r, f) - n(r, f') + n(r, 1/f') = \sum_{\zeta \in D_r} \nu_\zeta(f),$$

where

$$\nu_\zeta(f) = 2 \max\{\mathrm{ord}_\zeta(1/f), 0\} - \max\{\mathrm{ord}_\zeta(1/f'), 0\} + \max\{\mathrm{ord}_\zeta(f'), 0\}.$$

Note that if f has a pole of order p at $z = \zeta$ then $\nu_\zeta(f) = p - 1$. Otherwise $\nu_\zeta(f) = 0$. Now suppose that $f(\zeta) = a$ for some $a \in \mathbb{C}$. If f takes the value a with multiplicity p at $z = \zeta$ (i.e. if $f(z) = a + \alpha(z - \zeta)^p + O((z - \zeta)^{p+1})$, where $\alpha \neq 0$) then again $\nu_\zeta(f) = p - 1$. So $n_{\mathrm{ram}}(r, f)$ counts higher-multiplicity (or ramified) values. The corresponding enumerative function $N_{\mathrm{ram}}(r, f)$ is called the *ramification term*.

Theorem 4.1 (Nevanlinna's Second Main Theorem). *Let f be a non-constant meromorphic function. For $q \geq 2$, let $a_1, \ldots, a_q \in \mathbb{C}$ be q distinct points. Then*

$$(q - 1)T(r, f) \leq N(r, f) + \sum_{j=1}^{q} N\left(r, \frac{1}{f - a_j}\right) - N_{\mathrm{ram}}(r, f) + S(r, f),$$

where $N_{\mathrm{ram}}(r, f) = 2N(r, f) - N(r, f') + N(r, 1/f')$.

Using the ramification term, we have the following immediate corollary.

Corollary 4.2. *Let f be a non-constant meromorphic function. For $q \geq 2$, let $a_1, \ldots, a_q \in \mathbb{C}$ be q distinct points. Then*

$$(q - 1)T(r, f) \leq \bar{N}(r, f) + \sum_{j=1}^{q} \bar{N}\left(r, \frac{1}{f - a_j}\right) + S(r, f),$$

where $\bar{N}(r, f)$ counts poles ignoring multiplicities.

Corollary 4.3 (Picard's Theorem). *Let f be a transcendental meromorphic function which takes each of three distinct values in $\mathbb{C} \cup \{\infty\}$ at most finitely many times. Then f is a constant.*

Proof of Theorem 4.1. We will present a proof that gives a very good expression for the "error term" $S(r, f)$ if required.

Let

$$l = \min_{1 \le i < j \le q} |a_i - a_j|,$$

which is independent of f. Fix $z \in \mathbb{C}$ such that $f'(z) \ne 0$, $f(z) \ne \infty$ and $f(z) \ne a_j$ for all $j \in \{1, 2, \ldots, q\}$. Choose $k \in \{1, 2, \ldots, q\}$ such that $|f(z) - a_k| \le |f(z) - a_j|$, for all $j \in \{1, 2, \ldots, q\}$. For all $j \in \{1, 2, \ldots, q\} \setminus \{k\}$, we have

$$l \le |a_k - a_j| \le |f(z) - a_k| + |f(z) - a_j| \le 2|f(z) - a_j|.$$

Hence $|f(z) - a_j| \ge l/2$. This gives

$$\log^+ |f(z) - a_j| = \log |f(z) - a_j| + \log^+ \frac{1}{|f(z) - a_j|}$$

$$\le \log |f(z) - a_j| + \log^+ \frac{2}{l}.$$

So

$$\log^+ |f(z)| \le \log^+ |f(z) - a_j| + \log^+ |a_j| + \log 2$$

$$\le \log |f(z) - a_j| + \log^+ \frac{2}{l} + \log^+ |a_j| + \log 2.$$

Therefore

$$(q - 1) \log^+ |f(z)| \le \sum_{\substack{j=1 \\ j \ne k}}^{q} \log |f(z) - a_j| + C(a_1, a_2, \ldots, a_q), \qquad (4.1)$$

where

$$C(a_1, a_2, \ldots, a_q) = \log^+ \sum_{\substack{j=1 \\ j \ne k}}^{q} |a_j| + (q - 1)\left(\log^+ \frac{2}{l} + \log 2\right).$$

Now

$$\sum_{\substack{j=1 \\ j \ne k}}^{q} \log |f(z) - a_j| = \sum_{j=1}^{q} \log |f(z) - a_j| + \log \frac{1}{|f(z) - a_k|}$$

$$= \sum_{j=1}^{q} \log |f(z) - a_j| - \log |f'(z)| + \log \frac{|f'(z)|}{|f(z) - a_k|}$$

$$\le \sum_{j=1}^{q} \log |f(z) - a_j| - \log |f'(z)|$$

$$+ \log \left(\sum_{j=1}^{q} \frac{|f'(z)|}{|f(z) - a_j|} \right). \qquad (4.2)$$

The point of the last step is that none of the a_j's are privileged (in particular, a_k is not treated any differently from the others). Using the inequalities (4.1) and (4.2), we have

$$
(q-1)m(r,f)
$$

$$
= \frac{1}{2\pi} \int_0^{2\pi} (q-1) \log^+ |f(re^{i\theta})| d\theta
$$

$$
\leq \frac{1}{2\pi} \sum_{j=1}^q \int_0^{2\pi} \log |f(re^{i\theta}) - a_j| - \frac{1}{2\pi} \int_0^{2\pi} \log |f'(re^{i\theta})| d\theta
$$

$$
+ \frac{1}{2\pi} \int_0^{2\pi} \log \left(\sum_{j=1}^q \frac{|f'(re^{i\theta})|}{|f(re^{i\theta}) - a_j|} \right) d\theta + C(a_1, a_2, \ldots, a_q). \quad (4.3)
$$

Applying Jensen's formula (2.4) to $f'(z)$ and $f(z) - a_j$ to eliminate the first $q+1$ integrals in the inequality (4.3) gives

$$
(q-1)m(r,f) - \sum_{j=1}^q N\left(r, \frac{1}{f-a_n}\right) + qN(r,f) + N\left(r, \frac{1}{f'}\right) - N(r,f')
$$

$$
\leq \frac{1}{2\pi} \log \left(\sum_{j=1}^q \frac{|f'(re^{i\theta})|}{|f(re^{i\theta}) - a_j|} \right) d\theta + \sum_{j=1}^q \log |\mathrm{ilc}_0(f - a_j)| - \log |\mathrm{ilc}_0(f')|
$$

$$
+ C(a_1, a_2, \ldots, a_q).
$$

The left-hand side of this equation is

$$
(q-1)T(r,f) - \left\{ N(r,f) + \sum_{j=1}^q N\left(r, \frac{1}{f-a_n}\right) \right\} + N_{\mathrm{ram}}(r,f).
$$

To prove the theorem as stated, we note that

$$
\frac{1}{2\pi} \log \left(\sum_{j=1}^q \frac{|f'(re^{i\theta})|}{|f(re^{i\theta}) - a_j|} \right) d\theta \leq \frac{1}{2\pi} \int_0^{2\pi} \log^+ \left(\sum_{j=1}^q \frac{|f'(re^{i\theta})|}{|f(re^{i\theta}) - a_j|} \right) d\theta
$$

$$
\leq \sum_{j=1}^q m\left(r, \frac{f'(z)}{f(z) - a_j}\right) + \log q = S(r,f).
$$

Better estimates for the error term can be obtained using more subtle estimates for this term (see e.g. [2]).   □

4.2. Defect relations

For any $a \in \mathbb{CP}^1$, we introduce the notation

$$N(r, a, f) := \begin{cases} N(r, f) & \text{if } a = \infty, \\ N\left(r, \dfrac{1}{f - a}\right) & \text{if } a \in \mathbb{C}, \end{cases}$$

with similar definitions holding for $m(r, a, f)$, $n(r, a, f)$ etc. Recall that

$$T(r, a, f) = m(r, a, f) + N(r, a, f).$$

The defect relations are simple consequences of Nevanlinna's Second Main Theorem. They quantify the fact that, for most values of a, $m(r, a, f)$ is much smaller than $N(r, a, f)$.

For any non-constant meromorphic function f and all $a \in \mathbb{CP}^1$, we define the *Nevanlinna defect* to be

$$\delta(f, a) = \liminf_{r \to \infty} \frac{m(r, a, f)}{T(r, f)}.$$

Nevanlinna's First Main Theorem also gives the expression

$$\delta(f, a) = \liminf_{r \to \infty} \left(1 - \frac{N(r, a, f)}{T(r, f)}\right).$$

Clearly $0 \le \delta(f, a) \le 1$. A value $a \in \mathbb{CP}^1$ is said to be *deficient* for f if $\delta(f, a) > 0$ and *maximally deficient* if $\delta(f, a) = 1$. Roughly speaking, a deficient value is one that f takes fewer times than the maximum allowed by the growth of $T(r, f)$. In particular, any omitted value is maximally deficient.

We also introduce the *ramification defect*,

$$\theta(f, a) = \liminf_{r \to \infty} \frac{N(r, a, f) - \bar{N}(r, a, f)}{T(r, f)}.$$

and the *truncated defect* (to order 1),

$$\delta^{(1)}(f, a) := \delta(f, a) + \theta(f, a) = \liminf_{r \to \infty} \left(1 - \frac{\bar{N}(r, a, f)}{T(r, f)}\right).$$

Again, $0 \le \theta(f, a) \le 1$ and $0 \le \delta^{(1)}(f, a) \le 1$.

Theorem 4.4. *For a non-constant meromorphic function f the set of values $a \in \mathbb{CP}^1$ for which $\delta^{(1)}(f,a) > 0$ is countable and*

$$\sum_{a \in \mathbb{CP}^1} \delta^{(1)}(f,a) = \sum_{a \in \mathbb{CP}^1} \{\delta(f,a) + \theta(f,a)\} \leq 2.$$

In particular, there are only countably many deficient values (i.e. $a \in \mathbb{CP}^1$ such that $\delta(f,a) > 0$).

Proof. From Corollary 4.2 and Nevanlinna's First Main Theorem, we have that for any $q \geq 3$ distinct points $a_1, \ldots, a_q \in \mathbb{CP}^1$,

$$(q-2)T(r,f) \leq \sum_{j=1}^{q} \bar{N}(r, a_j, f) + S(r,f). \tag{4.4}$$

Hence, choosing a sequence (r_n) that lies outside the exceptional set of $S(r,f)$, we have that

$$q - \sum_{j=1}^{q} \frac{\bar{N}(r_n, a_j, f)}{T(r,f)} \leq (2 + o(1)),$$

as $n \to \infty$, from which it follows that

$$\sum_{j=1}^{q} \delta^{(1)}(f, a_j) \leq 2. \tag{4.5}$$

For all positive integers k, let $A_k = \{a \in \mathbb{CP}^1 : \delta^{(1)}(f,a) \geq 1/k\}$. The inequality (4.5) shows that each A_k contains at most $2k$ elements. Furthermore,

$$A := \{a \in \mathbb{CP}^1 : \delta^{(1)}(f,a) > 0\} = \bigcup_{k=1}^{\infty} A_k.$$

It follows that A is countable. $\qquad\square$

The next corollary follows from the fact that if f is entire then $\delta^{(1)}(f, \infty) = 1$.

Corollary 4.5. *If f is entire then*

$$\sum_{a \in \mathbb{C}} \delta^{(1)}(f,a) \leq 1.$$

Suppose that f has some value a such that the equation $f(z) = a$ has only multiple roots. Then we call a a *totally ramified* value for f. For any such value, $\bar{N}(r, a, f) \le \frac{1}{2} N(r, a, f) \le \frac{1}{2} T(r, f) + O(1)$. So $\delta^{(1)}(f, a) \ge 1/2$. So from Theorem 4.4 we have the following theorem.

Theorem 4.6. *Any non-constant meromorphic function has at most four totally ramified values.*

Nevanlinna's inverse problem was to show that, given any sequences (δ_j), (θ_j) and (a_j), where $0 \le \delta_j \le 1$, $0 \le \theta_j \le 1$ and $a_j \in \mathbb{CP}^1$, there exists a meromorphic function f with $\delta(f, a_j) = \delta_j$, $\theta(f, a_j) = \theta_j$ and $\delta(f, a) = \theta(f, a) = 0$ for all $a \notin \{a_j\}$. This was solved by Drasin in 1976 (see, e.g., [3]).

4.3. *Exercises*

(1) Let f be a meromorphic function. Show that

$$\mathcal{S}(f) := \{g : g \text{ is meromorphic and } T(r, g) = S(r, f)\}$$

is a field.
(2) Show that Theorem 4.1 is equivalent to the following theorem.

Theorem 4.7. *Let f be a transcendental meromorphic function. For $q \ge 2$, let $a_1, \ldots, a_q \in \mathbb{C}$ be q distinct points. Then*

$$m(r, f) + \sum_{j=1}^{q} m\left(r, \frac{1}{f - a_n}\right) \le 2T(r, f) - N_{\mathrm{ram}}(r, f) + S(r, f).$$

(3) Derive the inequality (4.4).
(4) Show that if f is transcendental and entire then it can have at most two totally ramified values.
(5) Show that if f is transcendental, entire and never zero then it cannot have any totally ramified values.

5.   Further Reading

There are several good references for classical Nevanlinna theory. Nevanlinna's own book [7] is very clear and takes time to motivate the material. However, this book is about much more than Nevanlinna theory. The books by Hayman [5] and Goldberg and Ostrovskii [4] have been the standard references in the field for many decades (although the latter was only recently translated into English). Cherry and Ye [2] gives a very readable overview with an emphasis on obtaining a very good form for the error term (the "$S(r, f)$" in the Lemma on the Logarithmic Derivative and hence in Nevanlinna's Second Main Theorem) and its relationship with the size of the exceptional set E. One reason for obtaining better estimates of these error terms is motivated by and analogy between Nevanlinna theory and Diophantine approximation. This connection is described in [2] but is the central focus of Ru's book [9]. Vojta [10] constructed a "dictionary" to relate ideas and definitions in Nevanlinna theory to those in Diophantine approximation. This led him to propose the important Vojta conjectures in Diophantine approximation. In Vojta's dictionary, a meromorphic function corresponds to an infinite set of numbers from a number field. The Nevanlinna characteristic corresponds to the logarithmic height. Bombieri and Gubler's book [1] on heights contains a chapter on Nevanlinna theory and a chapter on the Vojta conjectures.

Applications of Nevanlinna theory to differential equations are described in [6]. Nevanlinna theory for functions of several complex variables is discussed in the recent book by Noguchi and Winkelmann [8].

Recent developments in the field include a number of remarkable papers by Yamanoi in which several important estimates in classical Nevanlinna theory have been derived using methods from several complex variables. In some cases these estimates have been used to prove long standing conjectures. We end with one such estimate from [11]. In the following, $N_1(r, a, f) = N(r, a, f) - \bar{N}(r, a, f)$ counts higher multiplicity a-points.

Theorem 5.1. *Let f be a transcendental meromorphic function. Let $a_1, \ldots, a_q$ be $q \geq 2$ distinct complex numbers and let $\epsilon > 0$. Then*

$$(q - 1)\bar{N}(r, \infty, f) + \sum_{j=1}^{q} N_1(r, a_j, f) \leq N(r, 0, f^{(q)}) + \epsilon T(r, f)$$

for all $r > \mathrm{e}$ outside a set $E \subset (\mathrm{e}, \infty)$ of logarithmic density 0.

References

[1] E. Bombieri and W. Gubler, *Heights in Diophantine Geometry.* Cambridge University Press, Cambridge (2006).

[2] W. Cherry and Z. Ye, *Nevanlinna's Theory of Value Distribution.* Springer-Verlag, Berlin (2001).

[3] D. Drasin, On Nevanlinna's inverse problem. *Complex Variables Theory Appl.* **37**, 123–143 (1998).

[4] A. A. Goldberg and I. V. Ostrovskii, *Value Distribution of Meromorphic Functions.* Translations of Mathematical Monographs, Vol. 236. American

Mathematical Society, Providence, RI (2008). Translated from the 1970 Russian original by Mikhail Ostrovskii, With an appendix by Alexandre Eremenko and James K. Langley.

[5] W. K. Hayman, *Meromorphic Functions*. Clarendon Press, Oxford (1964).

[6] I. Laine, *Nevanlinna Theory and Complex Differential Equations*. Walter de Gruyter & Co., Berlin (1993).

[7] R. Nevanlinna, *Analytic Functions*. Translated from the second German edition by Phillip Emig. Die Grundlehren der Mathematischen Wissenschaften, Band 162. Springer-Verlag, Berlin (1970).

[8] J. Noguchi and J. Winkelmann, *Nevanlinna Theory in Several Complex Variables and Diophantine Approximation*. Springer, Tokyo (2014).

[9] M. Ru, *Nevanlinna Theory and Its Relation to Diophantine Approximation*, Vol. 1052. World Scientific Publishing Co., Inc., Singapore (2001).

[10] P. Vojta, *Diophantine Approximations and Value Distribution Theory*. Springer-Verlag, Berlin (1987).

[11] K. Yamanoi, Zeros of higher derivatives of meromorphic functions in the complex plane. *Proc. London Math. Soc.* **106**, 703–780 (2013).

Solutions and Hints for Selected Exercise

Section 2

(3) If $fg+gh+hf$ has a pole at z_0 then the order of this pole is bounded by the sum of the orders of any poles of f, g and h at z_0. So $n(r, fg+gh+hf) \leq n(r, f) + n(r, g) + n(r, h)$, giving $N(r, fg + gh + hf) \leq N(r, f) + N(r, g) + N(r, h)$. Furthermore

$$\log^+ |fg + gh + hf| \leq \log^+ (3 \max\{|fg|, |gh|, |hf|\})$$

$$\leq \log 3 + \max\{\log^+ |f| + \log^+ |g|, \log^+ |g|$$

$$+ \log^+ |h|, \log^+ |h|| + \log^+ |f|\}$$

$$\leq \log 3 + \log^+ |f| + \log^+ |g| + \log^+ |h|.$$

Hence $m(r, fg + gh + hf) \leq m(r, f) + m(r, g) + m(r, h) + \log 3$.

(5) we have

$$T(r, a_0 + a_1 f + \cdots + a_d f^d)$$

$$\leq T(r, a_0) + T(r, f(a_1 + a_2 f + \cdots + a_d f^{d-1}))$$

$$\leq T(r, a_0) + T(r, f) + T(r, a_1 + a_2 f + \cdots + a_d f^{d-1}) + \log 2$$

$$= T(r, f) + T(r, a_1 + a_2 f + \cdots + a_d f^{d-1}) + O(\log r).$$

The result follows by induction.

Section 3

(1) The equation has no constant solutions, hence by Theorem 3.6, for every meromorphic solution y and every $a \in \mathbb{C}$,

$$m\left(r, \frac{1}{y-a}\right) = S(r,y).$$

Also, the second Painlevé equation is of the form (3.7) in Theorem 3.7, where $N = 2$, $P(z,y) = 2y$ and $Q(z,y) = y'' - zy' - \alpha$, which gives $m(r,y) = S(r,y)$.

(3) $T(r, \cosh z) = T\left(r, \frac{1}{2}\frac{(e^z)^2 - 1}{e^z}\right) = 2T(r, e^z) + S(r, e^z)$, from Theorem 3.4. Hence $T(r, \cosh z) = 2r/\pi + S(r, e^z)$, so the order is 1.

(4) Just replace the derivatives $f^{(j)}(z)$ in the statement and proof of Theorem 3.6 by $f(z + c_j)$ for constants c_j and add the condition that y has finite order. Substituting $y(z) \equiv a$ in the difference equation gives

$$2a(1 - a^2) = az + \gamma,$$

which has no solution, since a is a constant. The conclusion follows from the difference version of Theorem 3.6.

Section 4

(4) For any totally ramified value a, $\delta^{(1)}(f, a) \geq 1/2$. The conclusion then follows from Corollary 4.5.

(5) We have $\delta^{(1)}(f, \infty) = \delta^{(1)}(f, 0) = 1$. Now apply Theorem 4.4.

www.ingramcontent.com/pod-product-compliance
Lightning Source LLC
Chambersburg PA
CBHW050556190326
41458CB00007B/2061